# A First Course in Control System Design
## 2nd Edition

# RIVER PUBLISHERS SERIES IN AUTOMATION, CONTROL AND ROBOTICS

*Series Editors:*

**ISHWAR K. SETHI**
*Oakland University*
*USA*

**TAREK SOBH**
*University of Bridgeport*
*USA*

**QUAN MIN ZHU**
*University of the West of England*
*UK*

Indexing: All books published in this series are submitted to the Web of Science Book Citation Index (BkCI), to SCOPUS, to CrossRef and to Google Scholar for evaluation and indexing.

The "River Publishers Series in Automation, Control and Robotics" is a series of comprehensive academic and professional books which focus on the theory and applications of automation, control and robotics. The series focuses on topics ranging from the theory and use of control systems, automation engineering, robotics and intelligent machines.

Books published in the series include research monographs, edited volumes, handbooks and textbooks. The books provide professionals, researchers, educators, and advanced students in the field with an invaluable insight into the latest research and developments.

Topics covered in the series include, but are by no means restricted to the following:

- Robots and Intelligent Machines
- Robotics
- Control Systems
- Control Theory
- Automation Engineering

For a list of other books in this series, visit www.riverpublishers.com

# A First Course in Control System Design
## 2nd Edition

**Kamran Iqbal**

University of Arkansas
Little Rock, USA

**River Publishers**

*Published, sold and distributed by:*
River Publishers
Alsbjergvej 10
9260 Gistrup
Denmark

www.riverpublishers.com

ISBN: 978-87-7022-152-8 (Hardback)
        978-87-7022-151-1 (Ebook)

# Contents

# Foreword

Dr. Kamran Iqbal has assembled a very valuable and comprehensive book on introductory and applied control. The book contains many tools and much knowledge about the details of mankind's and engineers' everyday implements, and their control. The concepts, issues, challenges and alternatives are lucidly and precisely presented.

The material is very well motivated and compactly and pragmatically presented. Matlab programs and computer tools supplant the discussions and analytical tools in order to ease the derivations and the conclusions.

A curious and motivated engineer or student of control will have no difficulty following the material, the mathematical developments and derivations, models of physical systems and interdisciplinary issues. Many examples and exercises are added to enrich the imagination and creativity of the reader. Subjects of Transfer Function, State Variables, Root Locus apparatus, Sample Date systems, Pole Assignment are precisely described. A variety of assigned problems and questions help the reader to become completely aware of alternatives and be able to envision different solutions in his or her mind.

The book is equipped with ample explanations, guidance, tools and encouragement. It will be a valuable tool in all countries of the world with different educational, industrial and technical facilities and where engineers are called upon to provide.

**Hooshang Hemami**
Professor emeritus of electrical and computer engineering
at Ohio State University
April 2020

# Preface

The aim of the second edition of A First Course in Control System Design, similar to the first edition is to present model-based control system design in a lucid, understandable and approachable manner. The book has been written with the needs of undergraduates and beginning graduate students in multiple engineering disciplines and practicing engineers in mind. The second edition is organized into nine chapters; the first half of the book is devoted to analysis and the second half to the design of control systems. The book covers the design of controllers for analog and sampled-data systems described by transfer function and state variable models. The coverage is restricted to models of single-input single-output (SISO) systems. Examples from diverse engineering disciplines are introduced in the first chapter and carried forward in the later chapters. MATLAB and Control Systems Toolbox are extensively used for design; occasionally, Symbolic Math Toolbox is also used. MATLAB scripts for solutions to all book examples are provided.

Control systems, both natural and man made, are pervasive in our lives. Our homes have environmental controls. The appliances we use at home, such as the washing machine, microwave, etc. have embedded controllers. We fly in airplanes and drive automobiles, which make extensive use of control systems. Our body regulates essential functions like blood pressure, heart-beat, breathing, and insulin levels in blood, manifesting biological control systems. The cells in the body regulate our metabolism and energy production using nutrient levels and electrolytes. The postural stability of the body depends on regulating body's center of mass (CoM) over the base of support. Fine motor control enables the manipulation and locomotion tasks we undertake as part of daily living. We essentially perform the control function as we walk or drive a car, the control objective in both cases being to follow a desired course at a preferred speed.

The industrial revolution in the eighteenth century ushered in the age of machines that needed automatic controls. As a result, ingenious solutions to the control problems were developed. An early example involved the use of centrifugal flyball governor for throttle adjustment to regulate the speed of

**Figure 1**  A generic control system block diagram that includes the controller, process to be controlled, the actuator and the sensor. The output is fed back and compared with the reference signal in the comparator.

the steam engine that was essential to industrial progress. Though control technology quickly developed to solve practical problems, the theoretical understanding of the control systems and its underlying design process was developed later in mid-twentieth century. The post WWII era launched the space age that focused on the optimal design of control systems, and their implementation via the computing machines. The quest for boosting the industrial output through factory automation has enabled advancement in the industrial process control, and in the industrial robots that makes extensive use of feedback control systems. The growing automation in the past few decades has increased our reliance on control systems.

A control system aims at realizing a desired behavior at the output of a device or system (the plant) by manipulating its input through a controller. Feedback based on observation of the process output via sensing elements plays and important role in automatic control systems (Figure 1). In the feedback control systems, the controller monitors the difference between the desired and actual values of output variables, and adjusts the system inputs accordingly by employing various control schemes. The control objective, often, is to reduce the error to zero at a sufficiently fast rate and maintain it there. The desired output may be expressed as a set point, that is, a constant value that the controller will try to maintain at the output. Alternatively, in tracking systems, the objective is to track a time-varying reference input. An example of the latter is the control system used to make a drone-mounted camera follow a moving object of interest.

The control system design is invariably undertaken to achieve multiple objectives. The first and the foremost among them is the stability of the closed-loop system, as the system outputs affect the inputs in real-time. The next objective is the dynamic stability or the ability of the controller to damp out the output oscillations, characterized by the damping ratio of the dominant response modes. Further, the controller aims to improve the speed of response, that is reflected by the system bandwidth. The steady-state response of the closed-loop system, ideally, has a unity transfer function, i.e., the system operates with no steady-state errors. Next, the controller is required to curtail the effect of disturbance and noise inputs on system output. A final objective in the control design is to impart robustness, which implies an ability to maintain performance levels in the presence of disturbance inputs, as well as its ability to withstand parameter variations and certain unmodeled dynamics.

The controller designed for stability and performance may be of static or dynamic type. In certain cases, a static gain controller may be adequate to achieve a desired level of performance. An example is the automobile in cruise control, where the gas intake is adjusted to affect the selected speed. As the performance demands increase, a simple gain control is no longer adequate and a dynamic controller becomes necessary. A dynamic controller is a dynamic system in its own right, which generates a time varying controller output that translates into the plant input. For example, the variation in the gas pedal while driving an automobile in cruise control in response to the climb or descent condition represents a time-varying controller output. An alternate understanding of the dynamic controller is a frequency-selective filter that emphasizes certain frequency bands in preference to others.

The dynamic controllers are traditionally distinguished as of phase-lead or phase-lag type. These provide, respectively, improvements to the transient or the steady-state response of the system. The two designs can be combined when needed. In the contemporary design methods, the controller combines one or more of the three basic control modes: a proportional (P), an integral (I), and a derivative (D) mode. The resulting proportional-integral-derivative (PID) controller is a general-purpose controller that has the ability to meet many of the control objectives as defined above. The PID controller is robust against variations in plant parameters, and is popular in industrial control systems. In the traditional design, the controller is implemented using analog circuits build with operational amplifiers and resistive-capacitive (RC) networks. In contemporary control systems, the controller is implemented as a software routine on a computer, a microcontroller, a DSP chip, or a programmable logic controller (PLC).

A digital controller appropriate for computer implementation may be obtained via emulation of an existing analog controller design. At high enough sampling rates, digital approximation of the analog controller provides comparable performance to the original analog controller. Alternatively, the design of the digital controller can be based on the pulse transfer function of the plant, i.e., a transfer function obtained via $z$-transform that is valid at the sampling intervals. Computer implementation of a controller invariably

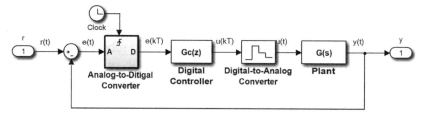

**Figure 2**   Block diagram of a digital control system that additionally includes an analog-to-digital converter (ADC) and a digital-to-analog converter (DAC).

adds phase lag to the feedback loop that compromises the stability margins. Hence, a more conservative controller design may be necessary if digital implementation of the controller is intended. This book addresses the controller design and their implementation for the analog systems as well as the sampled-data systems (Figure 2).

This book covers the control system design as applicable to single-input single-output (SISO) systems. The emphasis in this book is on understanding and applying the techniques that enable the design of effective control systems. The controller design is based on the mathematical model of the plant (the device or process to be controlled). System models are described in the frequency-domain using the transfer functions, or in the time domain using ordinary differential equations (ODEs). The state variable models describe the system in terms of time derivatives of a set of state variables. Control system design can be performed in either time or frequency-domain; essential design techniques for both are covered in this book. A limited number of skill assessment exercises are provided at the end of each chapter. Additional exercises can be found in standard control systems textbooks (listed as references at the end of the book).

The control systems concepts covered in this book are applicable to the various engineering disciplines. These concepts are typically covered at junior or senior level in the engineering curriculum. State variable models are covered in more depth in a beginning graduate course. Students in the scientific disciplines can also benefit from the control systems design concepts. A typical audience of this book includes inquisitive readers with interest in science, technology, engineering, and mathematics (STEM) fields. The mathematical background required for understanding, and hence benefitting from the material in this book, includes knowledge of linear algebra, complex numbers, and elementary differential equations. Additionally, some familiarity with Laplace and $z$-transforms is desired; the transform methods are reviewed in the Appendix.

The organization of this monograph is as follows: Chapter 1 discusses the modeling of physical systems. The dynamic character of such a system is typically captured using ODE models. Application of the Laplace transform converts the ODEs to algebraic equations in the Laplace transform variable '$s$', that are manipulated to obtain the input-output system description in the form of a transfer function (TF). The examples in this chapter include electrical, mechanical, electromechanical, thermal, and fluid systems. The chapter culminates with a discussion about linearizing the nonlinear dynamic system models.

Chapter 2 addresses the methods used to analyze the transfer function models. These models are characterized in terms of their poles and zeros. The poles effectively determine the modes of system natural response. The stability characterization requires the poles to be located in the open left half of the complex plane (OLHP). The system response to arbitrary inputs comprises natural and forced response components. The natural response of passive systems is of transitory nature and generally dies out with time. The forced response signifies the presence of a persistent input and is observable in the steady-state. System frequency response characterizes its response to sinusoidal inputs, that manifests as a sinusoid at the input frequency.

Chapter 3 addresses the methods of analysis for the state-variable models. State variables are often the natural variables, like inductor current and capacitor voltage in the electrical systems, or position and velocity of the inertial mass in the mechanical systems. The state variable description typically includes a set of matrices, that is, the system, the input, and the output matrices. A solution to the first-order state equations includes a convolution integral involving the state-transition matrix of the system. The choice of the state-variables for a given system is not unique, giving rise to several equivalent system descriptions, some of which may be preferred over others. The popular descriptions are the controller form, the observer form, and the modal form descriptions.

Chapter 4 introduces controller structures used with transfer function models. The static controller includes a scalar gain that multiplies the error signal generated by a comparator. The dynamic controllers include phase-lead, phase-lag and lead-lag types. Alternate description of the dynamic controllers includes the PD, PI, and PID controllers. Rate feedback controllers additionally make use of the rate signal and are similar to PD and PID controllers.

Chapter 5 discusses the control system design objectives. These include closed-loop stability, transient response improvement, steady-state error improvement, and improvement to the sensitivity robustness. The chapter discusses ways to characterize and achieve these objectives.

Chapter 6 discusses the root locus technique for designing cascade compensators for transfer function models. The root locus (RL) is the loci of roots of the closed-loop characteristic polynomial with variation in the controller gain. The RL technique primarily addresses the design of static controllers, but is easily extended to the design of first-order dynamic controllers. The chapter includes a discussion of the rate feedback design that may be preferred over cascade controller design. The chapter ends with a discussion of the controller realization methods with analog circuits using operational amplifiers and resister-capacitor networks.

Chapter 7 discusses the techniques to analyze and design sampled-data systems, that is, systems that include a clock-driven device, such as a microprocessor in the loop. The sampled-data systems are characterized by their pulse transfer function, that is obtained with the application of $z$-transform, and describes system input-output behavior at the sampling intervals. The time-domain description of sampled-data systems involves difference equations that can be solved by iteration. The stability of the sampled-data control system requires the poles of the closed-loop pulse characteristic polynomial to be located inside the unit circle in the complex $z$-plane. For high enough sampling rates, an approximate digital controller for sampled-data system can be obtained by emulating the analog controller designed for the continuous-time system. Alternatively, the pulse transfer function can be used to design a cascade digital controller by using the root locus technique.

Chapter 8 discusses the controller design for state variable models of analog and sampled-data systems. The state feedback design allows arbitrary placement of the roots of the closed-loop characteristic polynomial. The design is facilitated by first transforming the state variable model into the controller form. The pole placement method extends to the design of tracking systems that require placing an integrator in the feedback loop to achieve zero steady-state error. A continuous-time state variable model is converted to discrete-time by assuming a zero-order-hold at the input. The discrete-time state equations can be solved by iteration. A digital controller for the discrete state variable model can be similarly designed using the pole placement technique. Placing all poles of the closed-loop pulse transfer function at the origin results in a deadbeat design that guarantees the sampled-data system response to settle in $n$ time periods.

Chapter 9 discusses compensator design by frequency response modification. Frequency-domain methods, which predate the time-domain methods, utilize gain and phase margins to characterize the relative stability of the closed-loop system. The frequency-domain performance measures complement those used in time-domain analysis. A phase-lag compensator improves the DC gain and/or the phase margin of the system, resulting in steady-state or transient response improvement. A phase-lead compensator improves the system bandwidth hence its transient response. The two can be combined when both transient and steady-state response improvements are desired. The closed-loop frequency response can be visualized on the Nichol's Nyquist plot or the chart and reveals the presence of resonance in the system.

Throughout this book, symbols in regular font represent scalar variables, symbols in boldface letters represent arrays of variables, lower case letters represent vectors, and upper case letters represent matrices. Control systems designs presented in the book were performed in MATLAB (Mathworks, Inc.) using the Control Systems Toolbox; additionally Symbolic Math Toolbox and Simulink we used when needed. The MATLAB script used for system design and simulation is provided with the examples. Web links to solutions to the end-of-chapter problems are also provided. The figures representing control systems were mostly drawn in the Simulink GUI. Circuit models were drawn with TINA circuit simulator (Texas Instruments).

# Acknowledgement

I would like to extend my gratitude to my teachers for instilling the love of Control Systems in me.

I would like to thank my employer, University of Arkansas Little Rock, department of Systems Engineering for facilitating my work.

I would like to thank the editor and the publisher for helping me with this project.

I would like to thank my family and friends for their love and support.

I would like to thank my daughter Eeman for helping me draw the figures in Chapter 1 and with editing the manuscript.

# List of Figures

# List of Tables

# List of Abbreviations

| | |
|---|---|
| AC | Alternating current |
| BIBO | Bounded-input bounded-output |
| CL | Closed-loop |
| CLCP | Closed-loop characteristic polynomial |
| dB | Decibel |
| DC | Direct current |
| MIMO | Multi-input multi-output |
| OL | Open loop |
| OLHP | Open left half-plane |
| PD | Proportional-derivative |
| PFE | Partial fraction expansion |
| PI | Proportional-integral |
| PID | Proportional-integral-derivative |
| RL | Root locus |
| RLC | Resistor, inductor, and capacitor |
| SISO | Single-input single-output |
| ZOH | Zero-order hold |

## Common Symbols used in the book

| | |
|---|---|
| $s$ | Laplace transform variable |
| $z$ | Z transform variable |
| $G(s), G(z)$ | Plant transfer function |
| $K(s), K(z)$ | Controller/compensator transfer function |
| $H(s)$ | Sensor transfer function |
| $L(s)$ | Loop gain |
| $T(s), T(z)$ | Closed-loop transfer function |
| $r(t), r_k$ | Reference input |
| $u(t), u_k$ | Plant input |
| $y(t), y_k$ | Plant output |
| $e(t), e_k$ | Error signal |
| $d(t)$ | Disturbance input |
| $K$ | DC gain of the plant/controller |
| $\tau$ | Time constant |
| $\omega_n$ | Natural frequency |
| $\omega_d$ | Damped natural frequency |
| $\zeta$ | Damping ratio |
| $\boldsymbol{x}(t)$ | Vector of state variables |
| $\boldsymbol{x}_0$ | Vector of initial conditions |
| $\boldsymbol{A}, \boldsymbol{A}_d$ | System matrix in the state variable representation |
| $\boldsymbol{b}$ | Input distributions vector |
| $\boldsymbol{c}$ | Output contributions vector |
| $\boldsymbol{k}$ | Feedback gains vector |
| $e^{At}$ | State transition matrix |

# 1

# Mathematical Models
# of Physical Systems

## Learning Objectives

1. Obtain a differential equation model of a low order dynamic system.
2. Obtain system transfer function from the differential equation description.
3. Obtain a physical system model in the state variable form.
4. Linearize a nonlinear dynamic system model about an operating point.

In this chapter we discuss the process of obtaining mathematical description of a dynamic system, i.e., a system whose behavior changes over time. The models of continuous-time systems are primarily described in terms of linear or nonlinear differential equations.

Physical systems of interest to engineers include electrical, mechanical, electromechanical, thermal, and fluid systems, among others. Using the lumped parameter assumption, their behavior is mathematically described by ordinary differential equation (ODE) models. These equations are, in general, nonlinear, but can be linearized about an operating point for analysis and design purposes.

To model physical systems with interconnected components, individual component models can be assembled to obtain a system model. In the case of electrical circuits and networks, these elements include resistors, capacitors, and inductors. For mechanical systems, these include inertia (masses), springs, and dampers (or friction elements). For thermal systems, these include thermal capacitance and thermal resistance. For hydraulic and fluid systems, these include reservoir capacity and flow resistance.

The individual components or devices that form a physical system can store, exchange, or dissipate energy, which, gives rise to the time-varying or dynamic behavior of the system. The dynamic behavior is captured by the differential equation model. The natural variables associated with those elements, e.g., positions and velocities in mechanical systems, or currents and voltages in electrical systems, form the set of state variables commonly used to describe system behavior.

In physical systems, properties (or entities) may flow in and out of a system boundary, for example, heat, mass, or volume flowing in or out of a reservoir. To model such systems, conservation laws or balance equations maybe used to describe system dynamics in terms of rate of change of an accumulated property. Specifically, let $Q$ represent the accumulated property, and let $q_{in}$ and $q_{out}$ represent the inflow and outflow rates, then the relevant dynamic equation is described as:

$$\frac{dQ}{dt} = q_{in} - q_{out} + g - c$$

where $g$ and $c$ denote the internal generation and consumption.

The Laplace transform is commonly used to convert a set of linear differential equations into algebraic equations that can then be manipulated to obtain an input–output description, that is, a transfer function. The transfer function that describes the system behavior is a rational function of a complex frequency variable ($s$).

The transfer function forms the basis for analysis and design of control systems in the frequency domain. In contrast, the modern control theory is established on time-domain techniques involving the state equations that describe system behavior as time derivatives of a set of state variables. These are defined as any set of minimum number of variables that can capture system behavior under the various operating conditions.

Linearization of nonlinear models is accomplished by using Taylor series expansion about a critical point. The resulting linear model is only effective in the neighborhood of the critical or stationary point. The linear systems theory is well established and is relied upon for controller design, that is aimed to modify the system behavior.

In this chapter we discuss simples models of electrical, mechanical, electromechanical, thermal, and fluid systems that form the basis for system ananlysis and design techniques that are covered in the later chapters.

## 1.1 Modeling of Physical Systems

The mathematical modeling of a physical system is enabled by the choice of variables associated with the physical characteristics of its components.

These variables naturally divide into flow and across variables. The relationship between these two variables determines the element type, as discussed below.

### 1.1.1 Model Variables and Element Types

Modeling of a physical system involves two kinds of variables: flow variables that "flow" through the system components and across variables that are measured across those components. In electrical circuits, voltage or potential is measured across the circuit nodes, whereas current or electrical charge flows through the circuit branches.

In mechanical linkage systems, displacement and velocity are measured across the connecting nodes, whereas force or effort "flows" through the linkages. In the case of thermal and fluid systems, heat and mass serve as the flow variables, while temperature and pressure constitute the across variables.

The relationship between flow and across variables associated with an element in the system defines the type of physical element being modeled. The three basic types are the resistive, inductive, and the capacitive elements. The terminology that is specific to electrical circuits extends to many other types of physical systems.

To proceed further, let $q(t)$ denote a flow variable and $x(t)$ denote an across variable associated with a physical element; then, the element type is defined by their mutual relationships, as described hereunder.

**The resistive element.** The resistive element is described by the relation

$$x(t) = k \, q(t)$$

i.e., the flow and across variables for the element vary in proportion to each other.

As an example, the voltage and current relationship through a resistor is described by Ohm's law, $V(t) = Ri(t)$, that states that the voltage across a resistor varies in proportion to the current flowing through the resistor.

Similarly, the force–velocity relationship through a mechanical damper is described as $v(t) = \frac{1}{b} f(t)$ , i.e., the velocity increases in proportion to the applied force.

**The capacitive element.** The capacitive element is described by the relation

$$x(t) = k \int q(t) dt + x_0$$

i.e., the across variable varies in proportion to the accumulated amount of the flow variable. Alternatively, the flow variable varies proportionally with the rate of change of the across variable as $q(t) = \frac{1}{k}\frac{dx(t)}{dt}$.

As an example, the voltage and current relationship through a capacitor is given as $V(t) = \frac{1}{C}\int i(t)dt + V_0$, i.e., the voltage across the capacitor is proportional to the integral of current through it, where the current integral represents the accumulation of electrical charge; hence, $Q = CV$. The inverse relationship is described as $i(t) = C\frac{dV}{dt}$.

Similarly, the force–velocity relationship that governs an inertial mass element is given as $v(t) = \frac{1}{m}\int f(t)dt + v_0$. The inverse relation is the familiar Newton's second law of motion, $f(t) = m\frac{dv(t)}{dt} = ma(t)$, where $a(t)$ denotes acceleration.

**The inductive element.** The inductive element is described by the relation

$$x(t) = k\frac{dq(t)}{dt}$$

i.e., the flow variable is obtained by differentiating the across variable. Alternatively, the across variable varies in proportion to the accumulation of the flow variable.

As an example, the voltage–current relationship through an inductive coil in an electric circuit is given as $V(t) = L\frac{di(t)}{dt}$. The inverse relationship is described as $i(t) = \frac{1}{L}\int V(t)dt + i_0$. Similarly, the force–velocity relationship through a linear spring is given as $v(t) = \frac{1}{K}\frac{df(t)}{dt}$. The inverse relation is $f(t) = K\int v(t)dt + f_0$.

While the resistive element dissipates energy, both the capacitive and inductive elements store energy in some form. For example, a capacitor stores electrical energy and a moving mass stores kinetic energy. The energy storage accords memory to the element that accounts for the dynamic system behavior modeled by an ODE.

Next, we discuss system models obtained from the individual component descriptions. These models are described in increasing order of complexity, starting from the most basic ones.

### 1.1.2 First-Order ODE Models

Electrical, mechanical, thermal, and fluid systems that contain a single energy storage element are described by the first-order ODE models, where the mathematical model is conveniently described in terms of the output of the energy storage element. This is illustrated in the following examples.

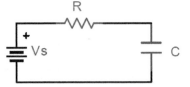

**Figure 1.1** An RC circuit.

**Figure 1.2** An RL circuit.

**Example 1.1:** A series RC network.

We consider a series RC network connected across a constant voltage source, $V_s$ (Figure 1.1). Kirchhoff's voltage law (KVL) is used to model the circuit behavior as $v_R + v_C = V_s$, where the capital letters are used to represent constant values and small letters represent time-varying quantities.

By substituting $v_C = v_0$, the ciruit output, and $v_R = iR = RC\frac{dv_0}{dt}$, we obtain the first-order ODE model that describes the circuit behavior as

$$RC\frac{dv_0(t)}{dt} + v_0(t) = V_s.$$

**Example 1.2:** A parallel RL network.

We similarly consider a parallel RL network connected across a constant current source, $I_s$ (Figure 1.2). The circuit is modeled by the first-order ODE, where the variable of interest is the inductor current, $i_L$, and Kirchhoff's current law (KCL) is applied at either of the nodes, to obtain $i_R + i_L = I_s$.

Then, by substituting $i_R = \frac{v}{R} = \frac{L}{R}\frac{di_L}{dt}$ we obtain the ODE dscription of the RL circuit as

$$\frac{L}{R}\frac{di_L(t)}{dt} + i_L(t) = I_s$$

We may note that the constant multiplier appearing with the derivative term in both RC and RL circuits defines the time constant of the circuit, i.e., the time when the system output in response to a constant input rises to 63.2% of its final value. The time constant is denoted by $\tau$ and is measured in [sec]. In particular, $\tau = RC$ for the RC circuit, and $\tau = L/R$ for the RL circuit.

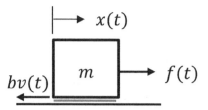

**Figure 1.3**  Motion of an inertial mass with surface friction.

**Example 1.3:** Inertial mass acted upon by a force.

The motion of an inertial mass, $m$, acted by a force, $f(t)$, in the presence of kinetic friction represented by $b$ is governed by Newton's second law of motion. The friction opposing motion is represented as $bv$, where $v$ is the velocity variable and $b$ is the friction constant. The resultant force on the mass element is $f - bv$. The resulting first-order ODE model for the system is given as

$$m\frac{dv(t)}{dt} + bv(t) = f(t).$$

The time constant for the mechanical model is $\tau = \frac{m}{b}$, and desribes the rate at which the velocity builds up in response to a constant force input.

**A Generic first-order ODE model.** The first-order ODE models in the above examples can be generalized. Accordingly, let $u(t)$ denote a generic input, $y(t)$ denote a generic output, and $\tau$ denote a time constant; then, the generic first-order ODE model is expressed as

$$\tau\frac{dy(t)}{dt} + y(t) = u(t).$$

Further examples of the first-order models include the thermal and fluid systems. These systems maybe modeled using balance equations applied to a control volume as illustrated hereunder.

**Example 1.4:** A model for room heating.

In order to model the room heating process, we assume that the heat flow into the room is denoted as $q_i$, the thermal capacity of the room is $C_r$, the temperature of the room is $\theta_r$, the ambient temperature is $\theta_a$, and the wall insulation is represented by a thermal resistance, $R_w$. Then, from the heat energy balance, we can write

$$C_r\frac{d\theta_r}{dt} + \frac{\theta_r - \theta_a}{R_w} = q_i.$$

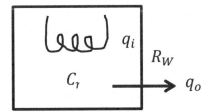

**Figure 1.4** Room heating with heatflow through walls.

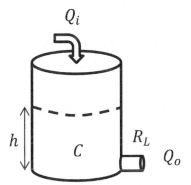

**Figure 1.5** Fluid reservoir with constricted outflow.

In terms of the temperature differential, $\Delta\theta = \theta_r - \theta_a$, the govering differential equation is

$$R_w C_r \frac{d\Delta\theta}{dt} + \Delta\theta = R_w q_i.$$

The temperature is measured in $[^\circ C]$, heat flow is measured in $[W]$, thermal capacitance is measured in $[\frac{J}{^\circ C}]$, and thermal resistance is measured in $[\frac{^\circ C}{W}]$.

We may note the similarity of the room heating model with the general first-order system model, $\tau\frac{dy}{dt} + y = u$, where the thermal time constant is given by $\tau = R_w C_r$.

**Example 1.5:** A model of hydraulic reservoir.

We consider a cylindrical reservoir filled with an incompressible fluid with a controlled exit at the bottom (Figure 1.5).

To proceed further, let $P$ denote the hydraulic pressure, $P_{atm}$ denote the atmospheric pressure, $A$ denote the area of the reservoir, $h$ denote the height, $V$ denote the volume, $\rho$ denote the mass density, $R_l$ denote the valve resistance to the fluid flow; $q_{in}$, $q_{out}$ denote the volumetric flow rates, and $g$ denote the gravitational constant. Then, the base pressure in the reservoir is obtained as

$$P = P_{atm} + \rho g h = P_{atm} + \frac{\rho g}{A} V.$$

Using reservoir capacitance, defined as: $C_h = \frac{dV}{dP} = \frac{A}{\rho g}$, the governing equation of the hydraulic flow through the reservoir is given as

$$C_h \frac{dP}{dt} = q_{in} - \frac{P - P_{atm}}{R_l}$$

In terms of the pressure difference, the equation is written as

$$R_l C_h \frac{d\Delta P}{dt} + \Delta P = R_l q_{in}.$$

Then, the above equation matches the standard first-order system model with $\tau = R_l C_h$. Further, using $\Delta P = \rho g h$ and $C_h = \frac{A}{\rho g}$, we can equivalently express the governing equation in terms of the liquid height, $h(t)$, in the reservoir as

$$A R_l \frac{dh}{dt} + \rho g h = R_l q_{in}$$

In the above, the hydraulic pressure is measured in $[\frac{N}{m^2}]$, volumetric flow is measured in $[\frac{m^3}{s}]$, hydraulic capacitance is measured in $[\frac{m^5}{N}]$, and flow resistance is measured in $[\frac{Ns}{m^5}]$.

### 1.1.3 Solving First-Order ODE Models with Step Input

The solution to a first-order ODE in the presence of a step forcing function and with given initial conditions can be obtained with the help of the Laplace transform (see Appendix). Accordingly, we consider the generic first-order ODE model,

$$\tau \frac{dy(t)}{dt} + y(t) = u(t).$$

We apply Laplace transform to the above equation assuming an initial condition, $y(0) = y_0$, to obtain:

$$\tau(sy(s) - y_0) + y(s) = u(s)$$

Next, assuming a unit step input $u(t)$, where $u(s) = \frac{1}{s}$, the output is solved as:

$$y(s) = \frac{1}{s(\tau s + 1)} + \frac{\tau y_0}{\tau s + 1}$$

We may use partial fraction expansion (PFE) to express the output as:

$$y(s) = \frac{1}{s} - \frac{\tau}{\tau s + 1} + \frac{\tau y_0}{\tau s + 1}$$

After applying the inverse Laplace transform, we obtain the time-domain solution to the first-order ODE as:

$$y(t) = [1 + (y_0 - 1)e^{-t/\tau}]u(t)$$

The $u(t)$ in the above expression represents a unit step function that is used to represent causality, i.e., the output is valid for $t \geq 0$.

**Steady-state output.** The steady-state value of the system response is denoted as $y_\infty = \lim_{t \to \infty} y(t)$. In terms of the steady-state output, the step response of the first-order ODE model is given as:

$$y(t) = [y_\infty + (y_0 - y_\infty)e^{-t/\tau}]u(t)$$

Assuming zero initial conditions, i.e., $y_0 = 0$, the output is expressed as:

$$y(t) = (1 - e^{-t/\tau})u(t)$$

We may evaluate the output at $t = k\tau, k = 0, 1, \ldots$ to compile the following table:

| Time | Output Value |
|------|--------------|
| 0 | $y(0) = 0$ |
| $1\tau$ | $1 - e^{-1} \cong 0.632$ |
| $2\tau$ | $1 - e^{-2} \cong 0.865$ |
| $3\tau$ | $1 - e^{-3} \cong 0.950$ |
| $4\tau$ | $1 - e^{-4} \cong 0.982$ |
| $5\tau$ | $1 - e^{-5} \cong 0.993$ |

By convention, the model output is assumed to have reached the steady state when the output attains 98% of its final value. Hence, the settling time of the system is given as $t_s = 4\tau$.

The output of the first-order ODE model to a unit-step input with zero initial conditions for $\tau = 1$ sec is plotted in Figure 1.6.

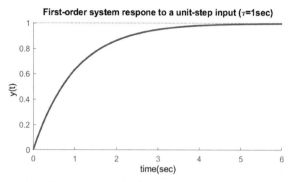

**Figure 1.6**   Output of a first-order system model to a unit-step input.

Further, the time variable in any first-order ODE model can be scaled by dividing it with the time constant. The response of a first-order ODE with nondimensionalized time variable $(t/\tau)$ is similar to that represented in Figure 1.6.

### 1.1.4 Second-Order ODE Models

A physical system that contains two energy storage elements is described by a second-order system model, where each energy-storing component contibutes a first-order term to the model. Examples of the second-order systems include RLC networks and the motion of an inertial mass with position output.

The modeling of a second-order system is illustrated by the following examples.

**Example 1.6:** Series RLC circuit.

A series RLC circuit with voltage input $V_s(t)$ and current output $i(t)$ has the following governing relationship obtained by applying KVL to the mesh:

$$L\frac{di(t)}{dt} + Ri(t) + \frac{1}{C}\int i(t)dt = V_s(t)$$

The integro-differential equation can by converted into a second-order ODE by expressing it in terms of the electric charge, $q(t)$, as

$$L\frac{d^2q(t)}{dt^2} + R\frac{dq(t)}{dt} + \frac{1}{C}q(t) = V_s(t)$$

Alternatively, the series RLC circuit behavior can be described in terms of the two first-order ODEs involving dual variables, the current, $i(t)$, and the

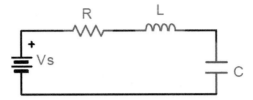

**Figure 1.7**   A series RLC circuit.

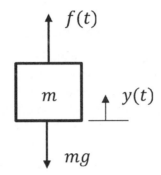

**Figure 1.8**   Motion of an inertial mass under gravity.

capacitor voltage, $V_c(t)$, as

$$L\frac{di(t)}{dt} + Ri(t) + V_c(t) = V_s(t),$$

$$C\frac{dV_c}{dt} = i(t)$$

The above formulation, where $i(t)$ and $V_c(t)$ serve as variables of interest, is known as the state variable formulation and is further discussed in Section 1.2.

**Example 1.7:** Inertial mass with position output.

An inertial mass moving in a constant gravitational field has both kinetic and potential energies and is modeled by a second-order ODE. For example, the vertical motion of a mass element of weight, mg, that is pulled upward by a force, $f(t)$, is described using position output, $y(t)$, by a second-order ODE:

$$m\frac{d^2y(t)}{dt^2} + mg = f(t).$$

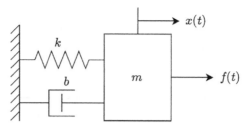

**Figure 1.9** A mass–spring–damper system.

**Example 1.8:** A mass–spring–damper system.

A mass–spring–damper system includes a mass affected by an applied force, $f(t)$, when its motion is restrained by a combination of a spring and a damper (Figure 1.9). Let $x(t)$ denote the displacement of the mass from a fixed reference; then, the dynamic equation of the system obtained by using Newton's second law of motion takes a familiar form, given as:

$$m\frac{d^2x(t)}{dt^2} + b\frac{dx(t)}{dt} + kx(t) = f(t).$$

The left-hand side in the above equation represents the sum of applied (inertial, damping, and spring) forces. In compact notation, we may express the ODE as:

$$m\ddot{x} + b\dot{x} + kx = f$$

where the dots above the variable represent time derivative, i.e., $\dot{x}(t) = \frac{dx(t)}{dt}$ and $\ddot{x}(t) = \frac{d^2x(t)}{dt^2}$.

In the absence of damping, the dynamic equation of the mass-spring system reduces to

$$m\frac{d^2x(t)}{dt^2} + kx(t) = f(t).$$

We may recognize that this equation models simple harmonic motion. Let $\omega_0^2 = k/m$; then, it can be verified by substitution that the general solution to the equation is given as:

$$x(t) = A\cos\omega_0\, t + B\sin\omega_0 t.$$

## 1.1.5 Solving Second-Order ODE Models

A second-order ODE model can be similarly solved by applying the Laplace transform to both sides of the differential equation. We consider a general second-order ODE model, described as:

$$\ddot{y}(t) + a_1\dot{y}(t) + a_2 y(t) = b_1\dot{u}(t) + b_0 u(t)$$

We note that the ODE inclues an input derivative term. We assume the following initial conditions for the ODE, i.e., $y(0) = y_0, \dot{y}(0) = \dot{y}_0, \dot{u}(0) = u_0$. Then, application of the Laplace transform gives an algebraic equation:

$$(s^2 + a_1 s + a_2)y(s) - (s + a_1)y_0 - \dot{y}_0 = (b_1 s + b_2)u(s) - b_1 u_0$$

which can then be solved for $y(s)$ to obtain

$$y(s) = \frac{1}{s^2 + a_1 s + a_2}[(s + a_1)y_0 + \dot{y}_0 + (b_1 s + b_2)u(s) - b_1 u_0]$$

The denominator in the above expression represents the characteristic polynomial of the second-order ODE model. Further, the characteristics of the time-response, $y(t)$, obtained from the application of inverse Laplace transform to $y(s)$, are determined by the roots of the characteristic equation: $s^2 + a_1 s + a_2 = 0$, as illustrated in the following example.

**Example 1.9:** The mass–spring–damper system.

We consider the mass-spring-damper system model (Example 1.8), in the presence of a unit-step input function, $u(t)$, assuming zero initial conditions; then, application of the Laplace transform produces the following algebraic equation:

$$(ms^2 + bs + k)x(s) = 2f(s)$$

Next, let $f(s) = 1/s$, and consider two cases based on the assumed system parameter values that results in either real or complex roots of the characteristic equation: $ms^2 + bs + k = 0$.

**Case I (real roots).** Let the parameter values be $m = 1, k = 2$, and $b = 3$; then, the characteristic equation is given as $s^2 + 3s + 2 = 0$. The equation has real roots at $s = -1, -2$.

The algebraic equation describing the system output is $(s^2+3s+2)x(s) = 2f(s)$. Assuming a unit-step input, $f(s) = \frac{1}{s}$, the output variable is solved as $x(s) = \frac{2}{s(s+1)(s+2)}$.

In order to apply the inverse Laplace transform, we need to carry out PFE of the output. We may use online SimboLab partial fraction calculator for this purpose (https://www.symbolab.com/solver/partial-fractions-calculator/). The result is given as

$$x(s) = \frac{1}{s} - \frac{2}{s+1} + \frac{1}{s+2}$$

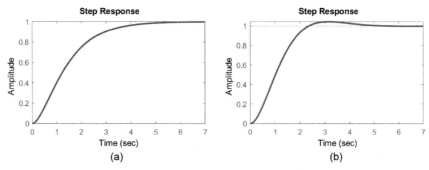

**Figure 1.10**   Time response of second-order system models: characteristic equation with real roots (a); with complex roots (b).

By applying the inverse Laplace transform, the time response of the spring–mass–damper system is obtained as (Figure 1.10(a))

$$x(t) = \left(1 - 2e^{-t} + e^{-2t}\right) u(t)$$

where $u(t)$ represents the unit-step function.

**Case II (complex roots).** Alternatively, let the parameter values be $m = 1$, $k = 2$, and $b = 2$; then, the characteristic equation is given as $s^2 + 2s + 2 = 0$. The equation has complex roots at $s = -1 \pm j1$.

The algebraic equation describing the system output is $(s^2 + 2s + 2)$ $x(s) = 2f(s)$. Assuming a unit-step input, the output variable is solved as $x(s) = \frac{2}{s(s^2 + 2s + 2)}$. Taking help from SimboLab for partial fractions, the output is given as

$$x(s) = \frac{1}{s} - \frac{s+2}{s^2 + 2s + 2}$$

The quadratic factor is alternately expressed as $(s+1)^2 + 1^2$, which helps when applying the inverse Laplace transform. Also, the quadratic term is split to write

$$x(s) = \frac{1}{s} - \frac{s+1}{(s+1)^2 + 1^2} - \frac{1}{(s+1)^2 + 1^2}$$

By applying the inverse Laplace transform, the time response of the spring–mass–damper system is obtained as (Figure 1.10(b)):

$$x(t) = (1 - e^{-t}\cos t - e^{-t}\sin t)u(t)$$

where $u(t)$ represents the unit-step function.

A comparison of the time response in the two cases reveals the following results:

1. The steady-state response in both cases settles at $x_\infty = G(0) = 1$.
2. The time response in the case of real roots resembles that of a first-order system with no overshoot.
3. The time response in the case of complex roots is oscillatory and has an overshoot.

## 1.2 Transfer Function Models

The transfer function model of a dynamic system is obtained by the application of the Laplace transform to the ODE model assuming zero initial conditions.

The transfer function describes the input–output relationship in the form of a rational function of a complex frequency variable "$s$."

We next describe the development of transfer function in the case of the first- and second-order systems.

**First-order ODE model.** We consider a generic first-order ODE model with input $u(t)$ and output $y(t)$, described as $\tau \frac{dy(t)}{dt} + y(t) = u(t)$. We use the Laplace transform to describe it as an algebraic equation, $(\tau s + 1)y(s) = u(s)$. The resulting input–output transfer function is given as

$$\frac{y(s)}{u(s)} = \frac{1}{\tau s + 1}$$

**Second-order ODE model.** We consider a mass–spring–damper model (Example 1.8), described by a second-order ODE, $m\ddot{x} + b\dot{x} + kx = f$. The model has a Laplace transform description:

$$ms^2 x(s) + bsx(s) + kx(s) = f(s).$$

The input–output relation (transfer function) for the mass–spring–damper system with force input and displacement output is given as

$$\frac{x(s)}{f(s)} = \frac{1}{ms^2 + bs + k}.$$

We may observe that the transfer function of a system is a ratio of two polynomials in $s$. We will assume, as is commonly the case, that the transfer function of a physical system is a proper fraction, i.e., the degree

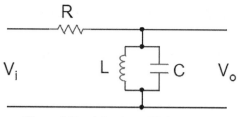

**Figure 1.11**   A bandpass RLC network.

of the denominator polynomial is greater than the degree of the numerator polynomial.

Higher-order models and multiple-input multiple-output (MIMO) system models described by linear ODEs can be similarly transformed into algebraic models and their transfer functions between designated input–output pairs can be defined.

**Example 1.10:** A bandpass RLC network.

We consider a bandpass RLC network with constant input voltage, $V_s$, and the output taken across the capacitor (Figure 1.11). By identifying the capacitor voltage $v_C$ and inductor current $i_L$ as natural variables, the two first-order ODEs that describe the network are given as

$$C\frac{dv_C}{dt} = \frac{V_s - v_C}{R} - i_L, \quad L\frac{di_L}{dt} = v_C$$

Application of Laplace transform then gives the corresponding algebraic equations as

$$\left(sC + \frac{1}{R}\right)v_C + i_L = \frac{V_s}{R}, \quad sLi_L - v_C = 0$$

A relationship between input $V_s$ and output $v_C$ is obtained by eliminating $i_L$ from the equations. The resulting transfer function is given as

$$\frac{v_C(s)}{V_s(s)} = \frac{sL/R}{s^2 LC + sL/R + 1}$$

### 1.2.1 DC Motor Model

A direct current (DC) motor (Figure 1.11) represents an electromechanical system that draws electrical energy and converts it into mechanical energy.

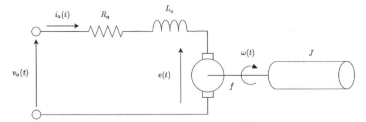

**Figure 1.12** An armature-controlled DC motor.

In an armature-controlled DC motor, the input is the armature voltage, $V_a(t)$, and the output is motor speed, $w(t)$, or the shaft angular position, $\theta(t)$.

In order to develop a model of the DC motor, let $i_a(t)$ denote the armature current, and $L$ and $R$ denote the electrical side inductance and the coil resistance, respectively. The mechanical side inertia and friction are denoted as $J$ and $b$, respectively. Let $k_t$ denote the torque constant and $k_b$ the motor constant; then, the dynamic equations of the DC motor for the electrical and mechanical sides are given as

$$L\frac{di_a(t)}{dt} + Ri_a(t) + k_bw(t) = V_a(t)$$

$$J\frac{dw(t)}{dt} + bw(t) - k_ti_a(t) = 0$$

By applying the Laplace transform, these equations are transformed into algebraic equations as

$$(Ls + R)i_a(s) + k_bw(s) = V_a(s)$$

$$(Js + b)w(s) - k_ti_a(s) = 0$$

In order to obtain a single input–output relation for the DC motor, we may solve the first equation for $i_a(s)$ and substitute in the second equation. Alternatively, we multiply the first equation by $k_t$, the second equation by $(Ls + R)$, and add them together to obtain

$$(Ls + R)(Js + b)w(s) + k_tk_bw(s) = k_tV_a(s)$$

From the above equation, the transfer function of the DC motor with voltage input and angular velocity output is derived as

$$\frac{w(s)}{V_a(s)} = \frac{k_t}{(Ls + R)(Js + b) + k_tk_b}$$

The angular position $\theta(s)$ is obtained by integrating the angular velocity $w(s)$; hence, the transfer function from $V_a(s)$ to the angular displacement $\theta(s)$ is given as

$$\frac{\theta(s)}{V_a(s)} = \frac{k_t}{s[(Ls + R)(Js + b) + k_t k_b]}$$

The denominator polynomial in the motor transfer function typically has real and distinct roots, which are reciprocal of the time constants contributed by the electrical and the mechanical sides of the motor $(\tau_e, \tau_m)$. In terms of the time constants, the DC motor model is alternatively described as

$$\frac{w(s)}{V_a(s)} = \frac{k_t/JL}{(s + 1/\tau_e)(s + 1/\tau_m)}$$

Of the two time constants, the electrical constant that represents the rate of build up of electrical current is relatively small compared to the mechanical constant that represents the build up of motor speed. Hence, the slower mechanical time constant dominates the overall motor response to a change in armature voltage, as shown in the following example.

**Example 1.11:** A DC motor model.

We assume that the parameter values for a small DC motor are given as $R = 1\Omega$, $L = 0.01H$, $J = 0.01$ kgm$^2$, $b = 0.1$ $\frac{\text{N–s}}{\text{rad}}$, and $k_t = k_b = 0.05$; then, the transfer function of the DC motor is obtained as

$$\frac{w(s)}{V_a(s)} = \frac{500}{(s + 100)(s + 10) + 25} = \frac{500}{(s + 10.28)(s + 99.72)}$$

The two time constants of the DC motor are given as $\tau_e \cong 10$ ms, $\tau_m \cong 100$ ms. We may note that $\tau_e$ matches the time constant of an RL circuit $(\tau_e = L/R)$ and $\tau_m$ matches the time constant of inertial mass in the presence of friction $(\tau_m = J/b)$.

Next, assuming a unit-step input, the output of the DC motor is given as

$$w(s) = \frac{500}{s\,(s + 10.28)\,(s + 99.72)} = \frac{0.488}{s} - \frac{0.544}{s + 10.28} + \frac{0.056}{s + 99.72}$$

By applying the inverse Laplace transform, the time-domain output is given as (Figure 1.13(a))

$$w(t) = [0.488 - 0.544e^{-10.28t} + 0.056e^{-99.72t}]u(t)$$

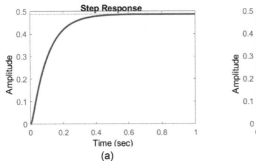 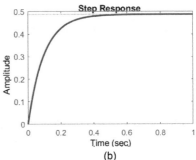

**Figure 1.13**    DC motor response to unit-step input: second-order motor model (a); first-order motor model (b).

**Simplified DC motor model.** The DC motor model developed earlier is a second-order ODE model with two unequal (electrical and mechanical) time constants, where $\tau_e \ll \tau_m$. Further, we observe that the motor response is dominated by the slower time constant $\tau_m$. Hence, we explore the possibility of describing the DC motor in terms of a first-order model with a single time constant.

A simplified model of the DC motor is obtained by ignoring the coil inductance ($L \to 0$), i.e., effectively ignoring the electrical time constant. The motor speed equation is modified as

$$R(Js + b)\omega(s) + k_t k_b \omega(s) = k_t V_a(s)$$

The resulting first-order DC motor transfer function is given as

$$\frac{\omega(s)}{V_a(s)} = \frac{k_t/R}{Js + b + k_t k_b/R}$$

The simplified model of a DC motor includes a single motor time constant

$$\left( \tau_m = \frac{JR}{bR + k_t k_b} \right),$$

and is given as

$$\frac{\omega(s)}{V_a(s)} = \frac{k_t/JR}{s + 1/\tau_m}$$

**Example 1.12:** A DC motor model (simplified).

Using the parameter values for a small DC motor (Example 1.11), its simplified transfer function model is obtained as

$$\frac{w(s)}{V_a(s)} = \frac{5}{s + 10.25}$$

The resulting motor time constant is $\tau_m \cong 97.6$ ms, which approximates the slower time constant in the second-order model. Assuming a unit-step input, the motor response is given as

$$w(s) = \frac{5}{s(s + 10.25)} = \frac{0.488}{s} - \frac{0.488}{s + 10.25}$$

By applying the inverse Laplace transform, the time-domain output is given as (Figure 1.13(b))

$$w(t) = \left[0.488 - 0.488e^{-10.25t}\right] u(t)$$

The motor response to unit-step input using the regular and simplified model is plotted and compared in Figure 1.13.

## 1.2.2 Industrial Process Models

Industrial processes comprise procedures involving exchange of chemical, electrical, or mechanical energy to aid in the manufacturing of industrial products. Industrial process models are mathematical models used to describe those processes.

An industrial process model, in its simplified form, can be represented by a first-order ODE accompanied by a dead time, i.e., there is a finite time delay between the application of the input and the appearance of the process output.

Let $\tau$ represent the time constant associated with an industrial process, $\tau_d$ represent the dead time, and $K$ represent the process DC gain; then, the simplified industrial process dynamics are represented by the following delay-differential equation:

$$\tau \frac{dy(t)}{dt} + y(t) = Ku(t - t_d).$$

Application of the Laplace transform produces the following first-order-plus-dead-time model of an industrial process:

$$G(s) = \frac{Ke^{-\tau_d s}}{\tau s + 1}.$$

The process parameters $\{K, \tau, \tau_d\}$, respectively, denote the process gain, the process time constant, and the process dead time. These parameters can be identified from the process response to a unit-step input.

We note that the process model involves a delay term represented by a transendental function, $e^{-\tau_d s}$. For analysis and controller design purposes, this term is replaced by a suitable rational approximation. Typical approximations obtained from Taylor series expansion of the transcendental term are given as

$$e^{-\tau_d s} \approx 1 - \tau_d s, \, e^{-\tau_d s} = \frac{1}{e^{\tau_d s}} \approx \frac{1}{1 + \tau_d s},$$

$$e^{-\tau_d s} = \frac{e^{-\tau_d s/2}}{e^{\tau_d s/2}} \approx \frac{1 - \tau_d s/2}{1 + \tau_d s/2}.$$

The latter expression above is called the first-order Padé approximation and is popular in industrial process control applications.

**Example 1.13:** A bioreactor model.

The process parameters of a stirred-tank bioreactor are given as $\{K, \tau, \tau_d\} = \{20, 0.5, 1\}$. The transfer function model of the process is formed as $G(s) = \frac{20e^{-s}}{0.5s+1}$.

By using a first-order Pade' approximation, a rational transfer function model of the industrial process with delay is obtained as: $G(s) = \frac{20(1-0.5s)}{(0.5s+1)^2}$.

We note that the above model has nonminimum phase due to the extra phase added by the right-half-plane zero in the numerator (see Chapter 2 for a description of poles and zeros).

## 1.3 State Variable Models

State variable models are time-domain models that express system behavior as time derivatives of a set of state variables. The state variables are often the natural variables associated with the energy storage elements appearing the system. The system order equals the number of such elements in the system.

In the case of electrical circuits, capacitor voltage and inductor currents serve as natural state variables. In the case of mechanical systems modeled with inertial elements, position and velocity of the inertial mass serve as natural state variables. In thermal systems, heat flow is a natural state variable. In hydraulic systems, the head (height of the liquid in the reservoir) is a natural state variable.

The state equations of the system model describe the time derivates of the state variables. When the state equations are linear, they are expressed in a vector-matrix form.

We note that the choice of state variables for a given system is not unique. Hence, equivalent state variable models of a physical system can be developed. All such models, however, share the same transfer function (see Chapter 3 for further discussion).

Examples 1.14–1.17 below illustrate the development of state-variable models in the case of the electrical, mechanical, and electromechanical systems.

**Example 1.14:** Series RLC circuit.

The governing equation of a series RLC circuit driven by a constant voltage source, $V_s$, with mesh current used as the circuit variable is given as (Example 1.6):

$$L\frac{di(t)}{dt} + Ri(t) + \frac{1}{C}\int i(t)dt = V_s$$

The circuit contains two energy storage elements: an inductor and a capacitor. Accordingly, let the inductor current, $i(t)$, and the capacitor voltage, $v_c(t)$, serve as state variables for the circuit. The state equations represent time derivatives of the state variables, expressed as

$$C\frac{dv_c}{dt} = i, \quad L\frac{di}{dt} = V_s - v_c - Ri.$$

We may note that the right-hand sides expressions in both equations contain state and input variables. In vector-matrix form, these equations are given as

$$\frac{d}{dt}\begin{bmatrix} v_c \\ i \end{bmatrix} = \begin{bmatrix} 0 & 1/C \\ -1/L & -R/L \end{bmatrix}\begin{bmatrix} v_c \\ i \end{bmatrix} + \begin{bmatrix} 0 \\ 1/L \end{bmatrix}V_s.$$

Let $v_c$ denote the RLC circuit output; then, an output equation is formed as

$$v_c = \begin{bmatrix} 1 & 0 \end{bmatrix}\begin{bmatrix} v_c \\ i \end{bmatrix}.$$

**Example 1.15:** The mass–spring–damper system.

The dynamic equation of the mass–spring–damper system is given as

$$m\frac{d^2x(t)}{dt^2} + b\frac{dx(t)}{dt} + kx(t) = f(t).$$

Let the mass position, $x(t)$, and the mass velocity, $v(t) = \dot{x}(t)$ serve as state variables, and let $x(t)$ be the output variable. The resulting state variable model of the mass–spring–damper system is given in terms of the state and output equations represented in matrix form as

$$\frac{d}{dt}\begin{bmatrix} x \\ v \end{bmatrix} = \begin{bmatrix} 0 & 1 \\ -k/m & -b/m \end{bmatrix} \begin{bmatrix} x \\ v \end{bmatrix} + \begin{bmatrix} 0 \\ 1/m \end{bmatrix} f$$

$$x = \begin{bmatrix} 1 & 0 \end{bmatrix} \begin{bmatrix} x \\ v \end{bmatrix}.$$

**Example 1.16:** The DC motor model.

The dynamic equations for the DC motor are given as

$$L\frac{di_a(t)}{dt} + Ri_a(t) + k_b\omega(t) = V_a(t).$$

$$J\frac{d\omega(t)}{dt} + b\omega(t) - k_t i_a(t) = 0.$$

Let $i_a(t)$, $\omega(t)$ serve as the state variables and let $\omega(t)$ be the output variable; then, the state variable model of the DC motor is given as

$$\frac{d}{dt}\begin{bmatrix} i_a \\ \omega \end{bmatrix} = \begin{bmatrix} -R/L & -k_b/L \\ k_t/J & -b/J \end{bmatrix} \begin{bmatrix} i_a \\ \omega \end{bmatrix} + \begin{bmatrix} 1/L \\ 0 \end{bmatrix} V_a.$$

$$\omega = \begin{bmatrix} 0 & 1 \end{bmatrix} \begin{bmatrix} i_a \\ \omega \end{bmatrix}.$$

Using the following parameter values for a small DC motor (Example 1.10): $R = 1\Omega, L = 1 \text{ mH}, J = 0.01 \text{ kgm}^2, b = 0.1\frac{\text{N·s}}{\text{rad}}, k_t = k_b = 0.05$, the state variable model of the motor is given as

$$\frac{d}{dt}\begin{bmatrix} i_a \\ \omega \end{bmatrix} = \begin{bmatrix} -100 & -5 \\ 5 & -10 \end{bmatrix} \begin{bmatrix} i_a \\ \omega \end{bmatrix} + \begin{bmatrix} 100 \\ 0 \end{bmatrix} V_a$$

$$\omega = \begin{bmatrix} 0 & 1 \end{bmatrix} \begin{bmatrix} i_a \\ \omega \end{bmatrix}.$$

**Example 1.17:** Bandpass RLC network.

We consider a bandpass RLC network (Example 1.10). The state variables for the network are selected as the capacitor voltage $v_C$ and inductor current $i_L$. The resulting state equations are given as

$$C\frac{dv_C}{dt} = \frac{V_s - v_C}{R} - i_L, \quad L\frac{di_L}{dt} = v_C$$

These equations are presented in the vector-matrix form as

$$\frac{d}{dt}\begin{bmatrix} v_c \\ i \end{bmatrix} = \begin{bmatrix} 0 & 1/C \\ -1/L & -R/L \end{bmatrix}\begin{bmatrix} v_c \\ i \end{bmatrix} + \begin{bmatrix} 0 \\ 1/L \end{bmatrix} V_s.$$

The output of the network is the capacitor voltage $v_C$. Hence, the output equation is given as

$$v_c = \begin{bmatrix} 1 & 0 \end{bmatrix}\begin{bmatrix} v_c \\ i \end{bmatrix}.$$

In state variable models, the choice of state variables affects the structure of the state variable representation for a given system. The choice itself is nonunique, so combinations of natural variables may also serve as state variables as long as the total number of variables stays the same. For example, we may use position and momentum in place of position and velocity as state variables in a mechanical system model (see Chapter 3 for further discussion).

## 1.4 Linearization of Nonlinear Models

In systems analysis, we come across both linear and nonlinear models of physical systems. The behavior of most physical systems is, in general, nonlinear, where the linear behavior can only be assumed for a limited range of operating conditions.

For example, the Ohm's law and the Hooke's law describe the linear behavior of a resistor and a mechanical spring. However, when considered over large dynamic input ranges, both components display nonlinear behavior.

The concept of linearity can be explained by expressing the system behavior as a mathematical function: $y = f(x)$. Let the input comprise a weighted sum of two components as $x = \alpha x_1 + \beta x_2$; then, a linear system model obeys the law of superposition:

$$f(\alpha x_1 + \beta x_2) = \alpha f(x_1) + \beta f(x_2).$$

Graphically, the input–output plot for a linear system is a straight line passing through the origin. The plot for nonlinear system may take any other shape and form. If the input–output behavior is a straight line that does not pass through origin, a simple change of variables can be employed to include the origin in the input–output graph.

In the case of general nonlinear systems, the linear behavior at a given operating point, $x = x_0$, maybe approximated by plotting a tangent line to the graph of $f(x)$ at that point. In the following, we generalize this notion to the models of dynamic systems.

### 1.4.1 Linearization About an Operating Point

Some physical system models are inherently nonlinear; an example is a simple pendulum model described by the dynamic equation:

$$ml^2\ddot{\theta}(t) + mgl \, \sin\theta(t) = T(t),$$

where $\theta(t)$ is the pendulum angle, $T(t)$ is the applied torque; $m, l$ represent the mass and the length of the pendulum respectively, and $g$ is the gravitational constant. Using $(\theta, \omega)$ as state variables, the nonlinear pendulum model is expressed in state variable form as

$$\frac{d}{dt}\begin{pmatrix}\theta\\\omega\end{pmatrix} = \begin{pmatrix}\omega\\-\dfrac{g}{l}\sin\theta\end{pmatrix} + \begin{pmatrix}0\\T(t)\end{pmatrix}.$$

As the Laplace transform is only applicable to linear models, in order to obtain a transfer function description, the nonlinear model of the pendulum first needs to be linearized.

Linearization of a nonlinear model involves first-order Taylor series expansion of the nonlinear function about a designated equilibrium point, i.e., the point where the time derivative is zero. The state variables for the linear model are selected as deviations relative to the equilibrium point.

Two equilibrium points are identified in the case of a simple pendulum: $\theta_e = 0°, 180°$. The linearized model of the pendulum involves first-order partial derivatives, expressed as a Jacobian matrix (a matrix of partial derivatives of a vector function with respect to a variable vector).

The Jacobian matrix for the simple pendulum is expressed as

$$\left[\frac{\partial f}{\partial x}\right] = \begin{pmatrix}0 & 1\\-\dfrac{g}{l}\cos\theta & 0\end{pmatrix}.$$

To formulate a linear model, the Jacobian is computed at a given equilibrium point. The linearized models, defined at the two equilibrium points: $\theta_e = 0°$, $180°$, are given as

$$\theta_e = 0°: \frac{d}{dt}\begin{bmatrix} \theta \\ \omega \end{bmatrix} = \begin{bmatrix} 0 & 1 \\ -\frac{g}{l} & 0 \end{bmatrix}\begin{bmatrix} \theta \\ \omega \end{bmatrix} + \begin{bmatrix} 0 \\ 1 \end{bmatrix} T(t)$$

$$\theta_e = 180°: \frac{d}{dt}\begin{bmatrix} \theta \\ \omega \end{bmatrix} = \begin{bmatrix} 0 & 1 \\ \frac{g}{l} & 0 \end{bmatrix}\begin{bmatrix} \theta \\ \omega \end{bmatrix} + \begin{bmatrix} 0 \\ 1 \end{bmatrix} T(t)$$

The output equation in both cases is given as $\theta(t) = \begin{bmatrix} 1 & 0 \end{bmatrix}\begin{bmatrix} \theta \\ \omega \end{bmatrix}$.

**Example 1.18:** A car driven under cruise control.

We consider a small car driven under normal conditions in cruise control. A block diagram of the cruise control system appears below (Figure 1.14).

The forces acting on the car include its weight, driving force generated by the engine torque, aerodynamic drag, and tire rolling friction.

We assume that the car weighs 1440 kg and is driven at 20 m/s (about 45 mph). The car experiences an aerodynamic drag force: $F_d = \frac{1}{2}\rho v^2 A c_d$. Assuming $\rho = 1.2$ kg/m$^3$ (for air), $A = 4$ m$^2$, and $c_d = .25$, results in a nonlinear drag force: $F_d = 0.6v^2 N$.

We assume that the tires generate a friction force: $F_r = 0.015W$, where $W$ is the weight of the car. For $m = 1440$ kg and $g = 9.8$ m/s$^2$, we obtain $F_r = 212N$.

Let $T_e = rF_e$ denote the engine torque, where $F_e$ denotes the force output and $r$ is the wheel radius; then, the dynamic equation of the car driven in cruise control is given as:

$$T_e/r - F_r - F_d = m\frac{dv}{dt}.$$

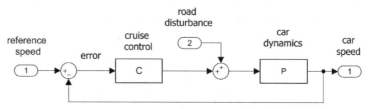

**Figure 1.14** A block diagram of the car cruise control system.

Substituting the parameter values including $r = 0.33$ m, we obtain:

$$m\frac{dv}{dt} + 0.6v^2 = 3T_e - 212.$$

We note that the above model is nonlinear. Next, we use first-order Taylor series expansion to obtain a linear approximation of the nonlinear term around our cruising speed (20 m/s). Thus,

$$F_d = F_d(20) + F_d'|_{v=20}(v - 20).$$

Let $\delta F_d = F_d - 20$, $\delta v = v - 20$, denote changes relative to the operating point; then, we have: $\delta F_d = 24\delta v$. Further, let $\delta T_e$ denote the change in the engine torque relative to the cruise conditions; then, the linearized model for the cruise control of a car traveling at 20 m/s is given as:

$$1440\frac{dv}{dt} + 24\delta v = 3\delta T_e.$$

## 1.4.2 Linearization of a General Nonlinear Model

We assume that the nonlinear state variable model of a single-input single-output (SISO) system is described by the following state and output equations:

$$\dot{\mathbf{x}}(t) = f(\mathbf{x}, u)$$

$$y(t) = g(\mathbf{x}, u),$$

In the above, $\mathbf{x}$ is a vector of state variables, $u$ is a scalar input variable, $y$ is a scalar output variable, $\mathbf{f}$ is a vector function of the state and input variables, and $g$ is a scalar function of those variables. Then, the linear state-space model is developed as follows.

Assume that a stationary point for the model is defined by: $f(\mathbf{x}_e, u_e) = 0$. Next, express the deviations from the stationary point as: $\mathbf{x}(t) = \mathbf{x}_e(t) + \delta\mathbf{x}(t)$; $u(t) = u_e(t) + \delta u(t)$. Then, using $\delta\mathbf{x}, \delta u$ as the new state and input variables, the linearized model is expressed as

$$\dot{\delta\mathbf{x}}(t) = [\partial f_i/\partial x_j]|_{(\mathbf{x}_e, u_e)}\delta\mathbf{x}(t) + [\partial f_i/\partial u]|_{(\mathbf{x}_e, u_e)}u(t)$$

$$y(t) = [\partial g/\partial x_j]|_{(\mathbf{x}_e, u_e)}\delta\mathbf{x}(t) + [\partial g/\partial u]|_{(\mathbf{x}_e, u_e)}\delta u(t),$$

where $[\partial f_i/\partial x_j]$ is a Jacobian matrix of partial derivatives; $[\partial f_i/\partial u]$, $[\partial g/\partial x_j]$ are vectors of partial derivatives, and $[\partial g/\partial u]$ is a scalar partial derivative, where all derivatives are computed at the stationary point.

The linearized model is expressed in its familiar vector-matrix form as

$$\dot{\delta}\mathbf{x}(t) = \mathbf{A}\delta\mathbf{x}(t) + \mathbf{b}u(t)$$
$$y(t) = \mathbf{c}^{\mathrm{T}}\delta\mathbf{x}(t) + d\delta u(t).$$

In the above description, $\mathbf{A}$ represents an $n \times n$ system matrix, $\mathbf{b}$ is a $n \times 1$ column vector of input distributions, $\mathbf{c}^{\mathrm{T}}$ is a $1 \times n$ row vector of output contributions, and $d$ is a scalar feed through term.

Finally, the above linearization process described for a SISO system can be generalized in the case of MIMO systems (see book references).

## Skill Assessment Questions

Link to the answers:
http://www.riverpublishers.com/book_details.php?book_id=449

1. Consider the first-order model of room heating, where the following parameter values are assumed: $C_r = 100[\frac{J}{°C}]$, $R_w = 0.2[\frac{°C}{W}]$, $\theta_a = 13°C$; assume that a $23°C$ room temperature is desired.

   (a) Formulate the room heating problem and compute the heat flow required to raise the room temperature by $10°C$.
   (b) Solve and sketch the room temperature $\theta(t)$ for a step change in the heat flow.

2. Consider a cylindrical hydraulic reservoir with a flow valve at the bottom with the following parameter values: $A = 1m^2$, $\rho = 1[\frac{m^3}{s}]$, $R_l = 50[\frac{Ns}{m^5}]$. Assume an input fluid flow of $q_{in} = 0.2\frac{m^3}{s}$, and let $g \cong 10\frac{m}{s^2}$.

   (a) Formulate the fluid flow problem.
   (b) Assuming that the reservoir was initially empty, solve for the steady-state height of the fluid in the reservoir.
   (c) Compute and sketch the height of the fluid $h(t)$.

3. Consider a series RLC circuit driven by a voltage source with capacitor voltage as the output variable.

   (a) Obtain a differential equation model for the circuit.
   (b) Obtain the input–output transfer function from the circuit model.
   (c) Define inductor current and capacitor voltage as state variables, and obtain a state variable description of the circuit model.

4. Consider a parallel RLC circuit driven by a current source with capacitor voltage as the output variable:

   (a) Obtain a differential equation model for the circuit.
   (b) Obtain the input–output transfer function from the circuit model.
   (c) Define inductor current and capacitor voltage as state variables, and obtain a state variable description of the circuit model.

5. Consider the model of a small DC motor, where the following parameter values are assumed: $R = 1\ \Omega$, $L = 10$ mH, $J = 0.01$ kgm$^2$, $b = 0.1\ \frac{Ns}{rad}$, $k_t = k_b = 0.02$.

   (a) Obtain the differential equations for the DC motor.

(b) Obtain the motor transfer functions with angular velocity and displacement as outputs.

(c) Obtain a state variable model of the motor with $\{i_a, \omega, \theta\}$ as state variables.

6. Consider the resonator circuit for a bandpass filter (Figure 1.11). Assume the following parameter values: $R = 1K$, $L = 10$ mH, $C = 1\mu F$; also, assume a unit-step input $V_s$. Compute and sketch the output $v_C(t)$.

7. Human postural dynamics in the sagittal plane are modeled as a rigid inverted pendulum described by

$$(I + ml^2)\ddot{\theta} - mgl \sin(\theta) = T$$

where $m$ is the mass of the body, $l$ is the vertical position of the center of mass, $g$ is the gravitational constant, $\theta$ is the angular displacement of the body from the vertical, and $T$ is stabilizing torque applied at the ankles. Assume the following parameter values: $m = 85$ kg, $l = 1$ m, $I = 15$ kg m$^2$, $g = 9.8$ m/s$^2$.

(a) Linearize the model about the vertical ($\theta = \dot{\theta} = 0$).

(b) Obtain the transfer function from ankle torque input to angular displacement output.

(c) Obtain a state variable description of postural dynamics with $\{\theta, \dot{\theta}\}$ selected as state variables.

8. The nonlinear model of an inverted pendulum over moving cart is described by the following equations:

$$(M + m)\ddot{y} + ml\ddot{\theta} \cos\theta - ml\dot{\theta}^2 \sin\theta = f$$

$$ml \cos\theta \, \ddot{y} + ml^2\ddot{\theta} - mgl \sin\theta = 0$$

where $m, l, \theta$ represent the mass, heigh, and angular position of the pendulum; and, $M, y$ represent the mass and position of the cart.

(a) Define state variables for the model and express the nonlinear model in the standard form: $\dot{x}(t) = f(x, u)$, $y = g(x)$, where $x = [y, \dot{y}, \theta, \dot{\theta}]$; $u = f$.

(b) Linearize the model about the equilibrium point: $\theta_e = 0, \dot{\theta} = 0$, $y = 0, \dot{y} = 0$.

# 2

---

# Analysis of Transfer Function Models

---

## Learning Objectives

1. Characterize a transfer function model by its natural response modes.
2. Determine the system response to step, impulse, and sinusoidal inputs.
3. Characterize the bounded-input bounded-output (BIBO) stability of the transfer function model.
4. Use Bode plot to visualize the frequency response of the system.

This chapter introduces analytical methods commonly applied toward the study, analysis, and design of transfer function models of physical systems. The transfer function characterizes the input–output behavior of a physical system and is obtained by the application of Laplace transform to the linear differential equation model of the system. The resulting transfer function, $G(s)$, is a rational function of a complex frequency variable, $s$. Further, given an input $u(s)$, the response of the system can be computed as: $y(s) = G(s)u(s)$.

The poles and zeros of the transfer function reveal important characteristics of the system response. The poles of the transfer function are roots of the denominator polynomial that determine the natural response modes of the system. These modes appear in the output whenever a system is excited through the application of input or through nonzero initial conditions.

A system may be characterized by its response to prototype inputs, such as unit-impulse and unit-step inputs. The system impulse response, that is, its response to a unit-impulse input with zero initial conditions, contains the natural modes of system response. The system step response, that is, its response to a unit-step input, comprises transient and steady-state components. The transient component contains natural modes of system response and usually dies out with time. The steady-state component persists and typically follows the forcing function.

Stability is a desired characteristic of the system. It refers to the system being well behaved and its output being predictable under various operating

31

conditions. The stability manifests through the transitory nature of the natural modes of system response, which die out with time in the case of stable systems.

The bounded-input bounded-output (BIBO) stability refers to the system response staying finite to every finite input. The BIBO stability requires that the poles of the system transfer function are located in the open left half of the complex $s$-plane.

The frequency response function, obtained by substituting $s = j\omega$ in the transfer function, characterizes the frequency response of the system, that is, its sinusoidal inputs. The response in the steady-state, that is, after the transient response has died out, is a sinusoid at the input frequency. Further, the magnitude of the system response is scaled by the gain of the system at the input frequency, and the output phase contains a component contributed by the system transfer function.

The frequency response of a system can be graphically visualized in various ways. We discuss the Bode plot representation in this chapter; other ways to visualize the frequency response are discussed in Chapter 9.

## 2.1 Characterization of Transfer Function Models

The transfer function, $G(s)$, of a system is a rational function in the Laplace transform variable, $s$. The transfer function relates the system response, $y(s)$, to an input, $u(s)$, that is, $y(s) = G(s)u(s)$. The transfer function is expressed as the ratio of the numerator and the denominator polynomials, that is, $G(s) = \frac{n(s)}{d(s)}$.

The main characteristics of transfer function models are discussed below.

### 2.1.1 System Poles and Zeros

Given a transfer function, $G(s)$, the roots of the numerator polynomial, that is, those frequencies at which the system response goes to zero, are called system zeros. Thus, $z_0$ is a zero of the transfer function if $G(z_0) = 0$.

The roots of the denominator polynomial define the system poles, that is, those frequencies at which the system response becomes infinite. Thus, $p_0$ is a pole of the transfer function if $G(p_0) = \infty$. The characteristic equation of the system is defined as: $d(s) = 0$.

The poles and zeros of the first- and second-order system models are characterized below.

**First-Order System.** A first-order system is typically described by the transfer function: $G(s) = \frac{K}{\tau s + 1}$; it has no finite zeros and one real pole at $s = -\frac{1}{\tau}$, where $\tau$ represents the time constant of the system.

**Second-Order System with a Pole at the Origin.** A second-order system with a pole at the origin is typically described by the transfer function: $G(s) = \frac{K}{s(\tau s+1)}$, where $\tau$ denotes the time constant. The system has no finite zeros and two poles at $s = 0$ and $s = -\frac{1}{\tau}$.

**Second-Order System with Real Poles.** The second-order system model of a DC motor is given as: $G(s) = \frac{K}{(s+1/\tau_e)(s+1/\tau_m)}$; this transfer function has no finite zeros and two real poles at $s_1 = -\frac{1}{\tau_m}$ and $s_2 = -\frac{1}{\tau_e}$, where $\tau_e$ and $\tau_m$ represent, respectively, the electrical and mechanical time constants.

**Second-Order System with Complex Poles.** The second-order system model of a spring–mass–damper system has the transfer function: $G(s) = \frac{1}{ms^2+bs+k}$. The characteristic equation is given as: $ms^s + bs + k = 0$. This transfer function has no finite zeros and two poles that are characterized by the discriminant, $\Delta = b^2 - 4\,mk$. Specifically,

- For $\Delta > 0$, the system has real poles, given as: $s_{1,2} = -\frac{b}{2m} \pm \sqrt{(\frac{b}{2m})^2 - \frac{k}{m}}$.
- For $\Delta < 0$, the system has complex poles, given as: $s_{1,2} = -\frac{b}{2m} \pm j\sqrt{\frac{k}{m} - (\frac{b}{2m})^2}$.
- For $\Delta = 0$, the system has two real and equal poles, given as: $s_{1,2} = -\frac{b}{2m}$.

The transfer function of a second-order system with a pair of complex poles may be represented as:

$$G(s) = \frac{K}{(s+\sigma)^2 + \omega_d^2}.$$

The transfer function poles are located at: $s_{1,2} = -\sigma \pm j\omega_d$, where $\sigma$ represents the real part of the pole, and $\omega_d$ refers to the damped natural frequency of the system, that is, the frequency of oscillation of the output response.

Equivalently, a second-order transfer function with complex poles is expressed as:

$$G(s) = \frac{K}{s^2 + 2\zeta\omega_n s + \omega_n^2},$$

where $\omega_n$ refers to the natural frequency of the system and $\zeta$ is the damping ratio. The transfer function poles are located at: $s_{1,2} = -\zeta\omega_n \pm j\omega_n\sqrt{1 - \zeta^2}$.

By comparing the two expressions for the transfer function with complex poles, we observe the following equivalence: $\sigma = \zeta\omega_n$ and $\sigma^2 + \omega_d^2 = \omega_n^2$.

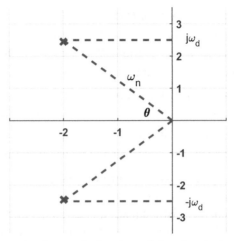

**Figure 2.1**   Second-order transfer function pole locations in the complex plane.

Further, $\omega_n$ equals the magnitude of the complex poles, and $\zeta = \frac{\sigma}{\omega_n} = \cos\theta$, where $\theta$ is the angle subtended by the complex poles at the origin (Figure 2.1).

**The Damping Ratio.** The damping ratio, $\zeta$, is a dimensionless quantity that characterizes the decay of the oscillations in the system's natural response. The damping ratio is bounded as: $0 < \zeta < 1$.

- As $\zeta \to 0$, the system poles are located close to the imaginary axis, and hence, oscillations in system's natural response persist for a long time, with $\omega_d \cong \omega_n$.
- As $\zeta \to 1$, the complex poles approach the real axis as $s_{1,2} \cong -\zeta\omega_n$; the high damping in the system eliminates oscillations in its natural response, so $\omega_d \to 0$. In the limiting case of $\zeta = 1$, we have the two poles co-located at: $s_{1,2} = -\zeta\omega_n$.

### 2.1.2 System Natural Response

The poles of the system transfer function characterize the natural modes of system response. If $p_i$ is a pole of the system transfer function, $G(s)$, then the corresponding natural response mode is described as $\{e^{p_i t}\}$.

The system transfer function may have real or complex poles. The real poles, denoted by $p_i = -\sigma$, contribute terms of the form $e^{-\sigma t}$ to the response. The complex poles, denoted by $p_i = -\sigma \pm j\omega$, contribute oscillatory terms of the form $e^{-\sigma t}e^{j\omega t}$ to the natural response.

The natural response of a system comprises a weighted sum of the natural modes of system response. The natural response modes are of transitory nature in the case of stable systems, that is, systems whose poles are located in the open left half of the complex plane, so that the natural response of the system dies out with time.

The natural modes of system response for the first- and second-order systems are described as follows:

**First-Order System.** The natural response of the first-order system has a single mode: $\{e^{-t/\tau}\}$, which is characterized by the time constant, $\tau$. Hence, the natural response is of the form:

$$y_n(t) = C_1 e^{-t/\tau},$$

where the arbitrary constant $C_1$ is determined by the initial conditions.

**Second-Order System with a Pole at the Origin.** The natural response of the second-order system with poles $\{0, -1/\tau\}$ has two modes, given as: $\{1, e^{-t/\tau}\}$. Hence, the natural response is of the form:

$$y_n(t) = C_1 + C_2 e^{-t/\tau},$$

where $\tau$ represents system time constant and coefficients $C_1$ and $C_2$ are determined by the initial conditions.

**Second-Order System with Real Poles.** The natural response of the second-order system with real and distinct poles, $s_{1,2} = -\sigma_1, -\sigma_2$, has two modes, given as: $\{e^{-\sigma_1 t}, e^{-\sigma_2 t}\}$. Hence, the natural response is represented as:

$$y_n(t) = C_1 e^{-\sigma_1 t} + C_2 e^{-\sigma_2 t}.$$

**Second-Order System with Complex Poles.** The response of the second-order system with complex poles, $s_{1,2} = -\sigma \pm j\omega$, has two modes, given as: $\{e^{(-\sigma \pm j\omega)t}\}$; an equivalent characterization of the natural response modes in this case is: $\{e^{-\sigma t} \cos \omega t, e^{-\sigma t} \sin \omega t\}$. Hence, the natural response is given as:

$$y_n(t) = (C_1 \cos \omega t + C_2 \sin \omega t)e^{-\sigma t}.$$

By using a trigonometric identity, we can alternately express the natural response as:

$$y_n(t) = C e^{-\sigma t} \sin(\omega t + \phi),$$

where $C = \sqrt{C_1^2 + C_2^2}$ and $\tan \phi = C_1/C_2$.

**System Time Constant.** A finite system pole $p_i$ contributes a mode $\{e^{p_i t}\}$ to the natural response of the system. The time constant associated with this mode is given as: $\tau = \frac{1}{p_i}$, which describes the rate of decay of the natural response mode: $\{e^{p_i t}\}$.

Specifically, the time constant describes the time when, starting from unity, the natural response decays to $e^{-1} \cong 0.37$, or 37% of its initial value. The effective time constant of a second-order system with complex poles is given as: $\tau_{eff} = \frac{1}{\sigma} = \frac{1}{\zeta \omega_n}$.

The oscillatory natural response of a system with complex poles is contained in the envelope defined by: $\pm e^{-\zeta \omega_n t}$.

## 2.2 System Response to Inputs

In this section, we consider the system response to prototype inputs that include a unit impulse function, $\delta(t)$, and the unit step function, $u(t)$. System response to sinusoidal inputs is considered later in Section 2.4.

### 2.2.1 The Impulse Response

The impulse response of a system is defined as its response to a unit-impulse input, $\delta(t)$. The impulse response exhibits inherent system properties, that is, those described by its natural response modes, and is used to characterize the system.

Assume that a dynamic system is described by its transfer function: $G(s)$; then, its response to an input, $u(t)$, is defined as: $y(s) = G(s)u(s)$.

In the case of a unit impulse input, $u(s) = 1$; hence, the impulse response is given as: $y(s) = G(s)$. The impulse response in the time-domain is computed as: $g(t) = L^{-1}[G(s)]$.

To explore further, we assume that system is described by a transfer function, represented as:

$$G(s) = \frac{n(s)}{\prod_{i=1}^{n} (s - p_i)}$$

where $n(s)$ is the numerator polynomial, and $p_i$, $i = 1, \ldots n$, are system poles that may include a single pole at the origin. For simplicity, we assume that the poles, $p_i$, are real and distinct. Then, using partial fraction expansion (PFE) with arbitrary constants, the impulse response is given as:

$$y_{imp}(s) = \sum_{i}^{n} \frac{A_i}{s - p_i}$$

Following the application of inverse Laplace transform, the impulse response of the system is computed as:

$$y_{imp}(t) = \sum_i^n A_i e^{p_i t}$$

We note that the impulse response is a weighted sum of the natural modes of system response. The impulse response dies out with time if the real parts of the poles obey: $Re(p_i) < 0$.

For systems with a simple pole at the origin, the impulse response approaches a constant value of unity in the steady-state. We may note that a pole at the origin represents an integrator and the integral of an impulse function is unity.

In particular, the impulse response of the first- and second-order systems is described below.

**First-Order System.** For a first-order system, let $G(s) = \frac{1}{\tau s+1}$; then, the impulse response is computed as:

$$g(t) = Ae^{-t/\tau}u(t),$$

where $A = \frac{1}{\tau}$. The $u(t)$ in the above expression represents the unit-step function, used here to indicate that the expression for $g(t)$ is valid for $t \geq 0$.

**Second-Order System with a Pole at the Origin.** For a second-order system with a pole at the origin, let: $G(s) = \frac{1}{s(\tau s+1)}$; then, using the PFE, $G(s) = \frac{1}{s} + \frac{\tau}{\tau s+1}$. After applying the inverse Laplace transform, the impulse response of the system is obtained as:

$$g(t) = (1 - e^{-t/\tau})u(t).$$

**Second-Order System with Real Poles.** For a second-order system with real poles, let $G(s) = \frac{K}{(s+a)(s+b)}$; then, using PFE followed by the inverse Laplace transform, the impulse response of the system is obtained as the following expression, where $A = \frac{1}{b-a}$:

$$g(t) = A(e^{-at} - e^{-bt})\,u(t),$$

**Second-Order System with Complex Poles.** For a second-order system with complex poles, let $G(s) = \frac{K}{(s+\sigma)^2+\omega_d^2}$; its impulse response is given as:

$$g(t) = Ae^{-\sigma t}\,\sin(\omega_d t)u(t),$$

where $A = \frac{K}{\omega_d}$. Alternatively, the impulse response of $\frac{Ks}{(s+\sigma)^2+\omega_d^2} =$ $\frac{K(s+\sigma)-K\sigma}{(s+\sigma)^2+\omega_d^2}$ is given as:

$$g(t) = Ke^{-\sigma t}\left(\cos\omega_d t + \frac{\sigma}{\omega_d}\sin\omega_d t\right)u(t)$$

Finally, a higher order system with dominant complex poles can be approximated and analyzed in terms of a typical second-order system.

**Example 2.1:** Unit-impulse response of first and second-order systems.

We compare the impulse response in the case of following transfer function models:

$$G_1(s) = \frac{1}{s+1}, G_2(s) = \frac{1}{s(s+1)}, G_3(s) = \frac{1}{(s+1)(s+2)}, G_4(s) = \frac{1}{s^2+2s+2}.$$

The impulse response in each case is plotted in MATLAB using the "impulse" command from the MATLAB Control Systems Toolbox (Figure 2.1). The MATLAB script is given below:

```
s=tf('s');                      % declare Laplace variable
G1=1/(s+1);                     % first-order transfer function
G2=1/(s*(s+1));                 % 2nd order TF, pole at the origin
G3=1/((s+1)*(s+2));             % 2nd order TF, real poles
G4=1/(s^2+2*s+2);              % 2nd order TF, complex poles
subplot(2,2,1), impulse(G1),grid % plot impulse response
subplot(2,2,2), impulse(G2),grid % plot impulse response
subplot(2,2,3), impulse(G3),grid % plot impulse response
subplot(2,2,4), impulse(G4),grid % plot impulse response
```

We make the following observations based on Figure 2.2:

1. The impulse response of a first-order system starts from unity. In contrast, the impulse response of a second-order system starts from zero.
2. The impulse response of a system model with poles in the open left-half plane, that is, with $Re[p_i] < 0$, approaches zero in the steady-state.
3. The impulse response of a system model with a pole at the origin approaches unity in the steady-state.

## 2.2.2 The Step Response

The step response of a system is defined as its response to a unit step input $u(t)$. Besides the system's natural response modes, the unit step response contains a constant term that emulates the forcing function, that is, the unit

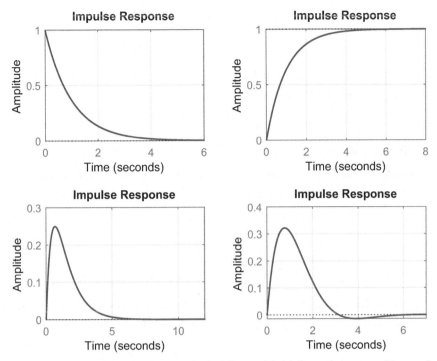

**Figure 2.2** Impulse response of a transfer function model: (a) first-order system; (b) second-order system with a pole at the origin; (c) second-order system with real poles; and (d) second-order system with complex poles.

step. System step response is often used to characterize the dynamic stability of the system model.

Let $G(s)$ describe the system transfer function; then, the unit-step response is given as: $y(s) = G(s)\frac{1}{s}$. The corresponding time-domain response is given as: $y(t) = \mathcal{L}^{-1}\left[\frac{G(s)}{s}\right]$.

The unit-step response of a system with zero initial conditions starts from zero, that is, $y(0) = 0$. The response settles down to a steady-state value $y_\infty = \lim_{t\to\infty} y(t)$, which may be determined from the application of the final value theorem as: $y_\infty = G(s)|_{s=0}$.

The unit step response in the case of the first- and second-order systems is described below.

**First-Order System.** For a first-order system, let $G(s) = \frac{K}{\tau s+1}$; then, its unit step response is computed as: $y(s) = \frac{K}{s(\tau s+1)} = \frac{K}{s} - \frac{\tau}{\tau s+1}$. For zero initial conditions, the time-domain response is given as:

$$y(t) = K(1 - e^{-t/\tau})u(t).$$

For arbitrary initial conditions, $y(0) = y_0$, the step response of a first-order system is given as:

$$y(t) = y_\infty + (y_0 - y_\infty)e^{-t/\tau}, \quad t \geq 0$$

where $y_\infty$ represents the unit step response in the steady-state.

**Second-Order System with a Pole at the Origin.** For a second-order system with a pole at the origin, let: $G(s) = \frac{K}{s(\tau s+1)}$; the step response is given as: $y(s) = \frac{K}{s^2(\tau s+1)}$. Using the PFE, we have: $y(s) = K(\frac{1}{s^2} - \frac{\tau}{s} + \frac{\tau^2}{\tau s+1})$; hence,

$$g(t) = K(t - \tau(1 - e^{-t/\tau}))u(t).$$

We may note that as $t \to \infty$, the response grows out of bound.

**Second-Order System with Real Poles.** For a second-order system with real poles, let $G(s) = \frac{K}{(\tau_1 s+1)(\tau_2 s+1)}$, $\tau_1 > \tau_2$; then, the step response is computed as:

$$y(s) = \frac{K}{s(\tau_1 s + 1)(\tau_2 s + 1)} = \frac{A}{s} + \frac{B}{\tau_1 s + 1} + \frac{C}{\tau_2 s + 1},$$

where $A = K, B = -\frac{K\tau_1^2}{\tau_1 - \tau_2}, C = \frac{K\tau_2^2}{\tau_2 - \tau_1}$. Hence,

$$y(t) = (A + Be^{-t/\tau_1} + Ce^{-t/\tau_2})u(t).$$

**Second-Order System with Complex Poles.** For a second-order system with complex poles, let $G(s) = \frac{K}{(s+\sigma)^2+\omega_d^2}$. Then, the unit-step response is computed as: $y(s) = \frac{A}{s} + \frac{Bs+C}{(s+\sigma)^2+\omega_d^2}$,

where $A = \frac{K}{\sigma^2+\omega_d^2}, B = -A, C = -2A\sigma$. Hence,

$$y(t) = \frac{K}{\sigma^2 + \omega_d^2}\left[1 - e^{-\sigma t}\left(\cos \omega_d t + \frac{\sigma}{\omega_d}\sin \omega_d t\right)\right]u(t).$$

Alternatively, we can use the phase form to express the unit step response as:

$$y(t) = \frac{K}{\sigma^2 + \omega_d^2}[1 - e^{-\sigma t}\cos(\omega_d t - \phi)]u(t), \quad \phi = \tan^{-1}\frac{\sigma}{\omega_d}.$$

**Example 2.2:** A small DC motor.

We consider the model of a small DC motor with the following parameter values: $R = 1\ \Omega$, $L = 10$ mH, $J = 0.01$ kg $-$ m$^2$, $b = 0.1\ \frac{\text{Ns}}{\text{rad}}$, $k_t = k_b = 0.05$; the DC motor input–output transfer function from armature voltage to motor speed is given as: $G(s) = \frac{500}{(s+10)(s+100)+25}$.

The step response of the DC motor is computed as:

$$y(t) \cong (0.49 - 0.54e^{-10.3t} + 0.05e^{-99.7t})u(t)$$

The DC motor electrical and mechanical time constants are: $\tau_e \cong 0.01\ s$ and $\tau_m \cong 0.1\ s$. Of these, the slower mechanical time constant dominates motor step response (Figure 2.3(a)).

In MATLAB (Control Systems Toolbox), the step response of the DC motor is plotted by using the following commands:

```
s=tf('s');              % declare Laplace variable
G=500/(s^2+110*s+1025); % DC motor transfer function
step(G), grid           % plot step response
```

**Example 2.3:** The mass–spring–damper system.

We consider the example of a mass–spring–damper system with the following parameter values: $m = 1$ kg, $b = 2\frac{Ns}{m}$, $k = 5\frac{N}{m}$. The resulting system transfer function is given as: $\frac{1}{s^2+2s+5} = \frac{1}{(s+1)^2+2^2}$, with complex poles located at: $s = -1 \pm j2$.

The unit-step response is computed as: $y(s) = \frac{1}{5}\left[\frac{1}{s} - \frac{s+2}{(s+1)^2+2^2}\right]$; then, by applying the inverse Laplace transform, we obtain the expression for system step response:

$$y(t) = \frac{1}{5}\left[1 - e^{-t}\left(\cos(2t) + \frac{1}{2}\sin(2t)\right)\right]u(t)$$

We observe that the step response of the system with complex poles is oscillatory (Figure 2.3(b)).

In MATLAB (Control Systems Toolbox), the step response of mass–spring–damper system is plotted by using the following commands:

```
s=tf('s');          % Laplace variable
G=1/(s^2+2*s+5);    % Mass-spring-damper system
step(G), grid       % plot step response
```

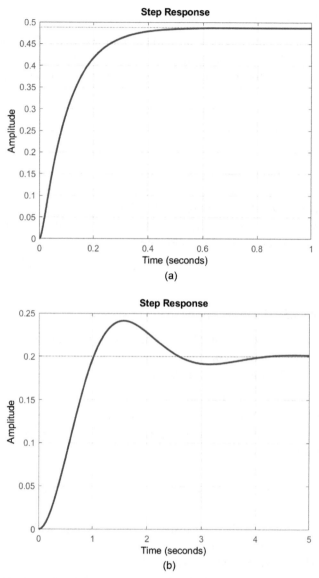

**Figure 2.3**   Step response of second-order systems: (a) DC motor model with real poles; (b) mass–spring–damper system with complex poles.

**Systems with Deadtime.** Models of industrial processes normally include a deadtime. The response of such systems starts after a delay, as illustrated in the following example.

**Example 2.4:** An industrial process model.

The simplified model of a stirred-tank bioreactor is given as: $G(s) = \frac{20e^{-s}}{0.5s+1}$. The step response of the bioreactor model is computed as: $y(s) = \frac{20e^{-s}}{s(0.5s+1)}$; hence,

$$y(t) = 20(1 - e^{-2(t-1)})u(t-1).$$

We may use the first-order Pade' approximation to obtain an approximate rational transfer function model as: $G_a(s) = \frac{20(1-0.5s)}{(0.5s+1)^2}$. The step response of the approximate bioreactor model, $G_a(s)$, is computed as: $y(s) = \frac{20(1-0.5s)}{s(0.5s+1)^2}$; hence,

$$y(t) = 20(1 - (1-4t)e^{-2t})u(t).$$

The two responses are compared below (Figure 2.4). The step response for the system with delay starts after the designated delay. The step response

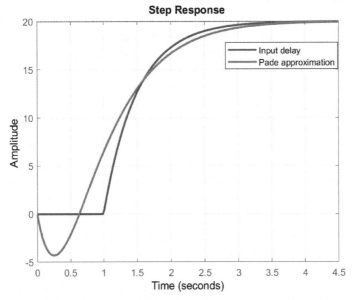

**Figure 2.4** Step response of an industrial process model with deadtime.

using Pade' approximation starts with an undershoot due to the presence of right-half-plane zero in the transfer function.

In the MATLAB (Control Systems Toolbox), the step responses are obtained by using the following commands:

```
s=tf('s');                        % declare Laplace variable
G=20/(.5*s+1);                    % define process model
Ga=G*(1-.5*s)/(1+.5*s);          % Pade' approx of delay
G.IODelay=1;                      % define process deadtime
step(G,Ga), grid                  % plot step response
legend('Input delay', 'Pade approximation')  % figure legend
```

We make the following observations based on Figure 2.4:

1. The step response of the process with deadtime starts after 1 s delay (as expected).
2. The step response of Pade' approximation of delay has an undershoot. This behavior is characteristic of transfer function models with zeros located in the right-half plane.

### 2.2.3 Characterizing the System Transient Response

The system response to an input in general, and unit-step input in particular, comprises two distinct components, a transient component that dies out with time and a steady-state component that persists, that is, $y(t) = y_{tr}(t) + y_{ss}(t)$.

The transient response component arises from the excitation of the natural response modes of the system; the steady-state response component arises solely because of the input and follows the input, that is, the forcing function.

The transient and steady-state components of system response for the first- and second-order systems are identified as follows:

- For a first-order system, $y_{tr}(t) = Be^{-t/\tau}u(t)$, and $y_{ss}(t) = A$.
- For a second-order system with real poles, such as a DC motor, we have: $y_{tr}(t) = (Be^{-t/\tau_m} + Ce^{-t/\tau_e})u(t)$ and $y_{ss}(t) = A$.
- For a second-order system with complex poles, for example, a spring–mass–damper system, we have: $y_{tr}(t) = (B\cos\omega_d t + C\sin\omega_d t)e^{-\sigma t}u(t)$ and $y_{ss}(t) = A$.

We note from the above first- and second-order system examples that the steady-state component of the step response is a constant, given as: $y_{ss}(t) = A$, where $A = G(s)|_{s=0}$.

Since the transient component of the step response of a stable system dies out with time, we may only be interested in its qualitative aspects. These are conveniently described using a proto-type second-order system with complex poles, as described below.

**Transient Response Quality Metrics.** The qualitative indicators of the unit-step response of a system with complex dominant poles and with zero initial conditions include the following:

1. The rise time ($t_r$), that is, time when the unit-step response first reaches unity,
2. The peak time ($t_p$), that is, when the unit-step response attains a peak value,
3. The response peak, $M_p$, and percentage overshoot (%$OS$),
4. The settling time ($t_s$), that is, the time when the response reaches and stays within 2% (alternately, 1%) of its steady-state value.

The above characteristics are conveniently illustrated by considering a prototype second-order system described by the transfer function:

$$G(s) = \frac{\omega_n^2}{s^2 + 2\zeta\omega_n s + \omega_n^2} = \frac{\omega_n^2}{(s + \zeta\omega_n)^2 + \omega_d^2}$$

where $\omega_n$ is the natural frequency of the system, that is, the frequency of oscillations in the absence of any damping; $\zeta$ is the damping ratio, and, $\omega_d = \omega_n\sqrt{1 - \zeta^2}$ is the damped natural frequency, that is, the frequency of oscillations in the actual system response.

The complex poles for the prototype system are located at: $s = -\zeta\omega_n \pm j\omega_d$. The impulse response of the prototype second-order system is given as:

$$y_{imp}(t) = \frac{\omega_n^2}{\omega_d}e^{-\zeta\omega_n t}\sin(\omega_d t)\, u(t)$$

The step response of the prototype system is given as:

$$y_{step}(t) = \left[1 - e^{-\zeta\omega_n t}\left(\cos(\omega_d t) + \frac{\zeta\omega_n}{\omega_d}\sin(\omega_d t)\right)\right]u(t)$$

The step response is alternatively expressed in amplitude-phase form as:

$$y_{step}(t) = \left[1 - \frac{e^{-\zeta\omega_n t}}{\sqrt{1 - \zeta^2}}\sin(\omega_d t - \phi)\right]u(t)$$

where $\tan\phi = \frac{\sqrt{1-\zeta^2}}{\zeta}$, and $\phi$ is the angle subtended to be the complex pole in the $s$-plane.

**Table 2.1**    Quality metrics used to measure second-order system response

| Quality Indicator | Expression |
|---|---|
| Rise time | $t_r = \dfrac{\pi - \phi}{\omega_d}, \quad \phi = \tan^{-1} \dfrac{\sqrt{1 - \zeta^2}}{\zeta}$ |
| Peak time | $t_p = \dfrac{\pi}{\omega_d}$ |
| Peak overshoot | $M_p = 100 e^{-\zeta \omega_n t_p}\,(\%)$ |
| Settling time | $t_s \cong \dfrac{4}{\zeta \omega_n} = \dfrac{4}{Re[p_i]}$ |

The rise time, $t_r$, when the step response has an overshoot is indicated when the step response first reaches unity, that is, when $\sin(\omega_d t_r - \phi) = 0$, so that $\omega_d t_r - \phi = \pi$; hence, the rise time is given as: $t_r = \frac{\pi - \phi}{\omega_d}$.

We note that $t_r$ is a function of $\zeta$. Specifically, $t_r$ reduces to $t_r = \frac{\pi}{2\omega_d}$ as $\zeta \to 0$, and increases to $t_r = \frac{\pi}{\omega_d}$ as $\zeta \to 1$. Hence, the rise time is bounded as: $\frac{\pi}{2\omega_d} \leq t_r \leq \frac{\pi}{\omega_d}$.

The peak time is indicated when the time derivative of the step response, or equivalently, when the impulse response approaches zero. The latter implies that $\sin(\omega_d t_p) = 0$; hence, $t_p = \frac{\pi}{\omega_d}$.

The peak overshoot in the step response is given as: $M_p = y_{step}(t_p) = 1 + e^{-\zeta \omega_n t_p}$; the percentage overshoot is given as: $\%OS = 100 e^{-\zeta \omega_n t_p}(\%)$.

The effective time constant of the prototype system response is given as: $\tau = \frac{1}{\zeta \omega_n}$. The response reaches and stays within 2% of its final value in about $4\tau$, and within 1% of its final value in $4.6\tau$; the settling time is given as $t_s \cong \frac{4}{\zeta \omega_n}$ (or $t_s \cong \frac{4.6}{\zeta \omega_n}$ when using 1% threshold).

Table 2.1 shows the qualitative indicators of the unit-step response in the case of prototype second order system.

**Example 2.5:** Prototype second-order system.

For a prototype second-order system, let $G(s) = \frac{10}{(s^2 + 4s + 10)}$. Then, the system poles are located at: $p_{1,2} = -2 \pm j2.45$, with $\omega_n = \sqrt{10}$ and $\zeta = \frac{2}{\sqrt{10}}$.

The step response of the system (Figure 2.5) displays a peak time of 1.29 s, a peak overshoot of 8%, and a settling time of 1.9 s (2% threshold).

## 2.2.4 System Stability

Stability is a desired characteristic of any dynamic system. Stability, in general terms, refers to the system being well behaved and in control under various operating conditions.

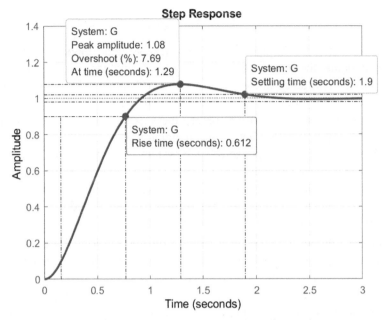

**Figure 2.5**   System step response with quality metrics.

Stability of a physical system is related to energy dissipation in the system. Stability is guaranteed in the case of systems built with passive components that either store or dissipate energy. Energy dissipation occurs due to friction, electrical resistance, thermal resistance, etc. Asymptotic stability implies that the residual energy in a system decays to zero over time.

Stability may be categorized in multiple ways, some of which are discussed below.

**Bounded-Input Bounded-Output Stability.** A popular stability measure in the case of transfer function models is that of bounded-input bounded-output (BIBO) stability, which implies that for every bounded input, $u(t): |u(t)| < M_1 < \infty$, the output of the system stays bounded, that is, $y(t): |y(t)| < M_2 < \infty$.

The output of a linear time-invariant (LTI) system is given by the convolution integral: $y(t) = \int_0^t g(t - \tau)u(\tau)d\tau$, where $g(t)$ denotes the impulse response. Thus, a necessary and sufficient condition for BIBO stability is that: $\int_0^\infty |g(t)|dt < \infty$, which requires that: $\lim_{t\to\infty} y_{tr}(t) = 0$.

The transient response of a linear system depends on its transfer function pole locations. In particular, $y_{tr}(t) = \sum_{i=1}^n A_i e^{p_i t}$, where $p_i$ is a pole of the transfer function. Hence, the requirement for BIBO stability requires that: $Re[p_i] < 0$.

We note that in the case of physical systems modeled with passive components, the transfer function commonly includes a damping term due to, that is, resistance in electrical systems or friction in mechanical systems. Hence, the condition $Re[p_i] < 0$ is duly satisfied.

A system model whose transfer functions have simple poles located on the stability boundary, that is, the imaginary axis, is termed as marginally stable. The impulse response of such systems does not go to zero, but stays bounded in the steady-state.

For example, a simple harmonic oscillator described by the ODE: $\ddot{y} + \omega_n^2 y = 0$, where $\omega_n$ represents the natural frequency, and the system has no damping. The oscillator transfer function is given as: $G(s) = \frac{1}{s^2 + \omega_n^2}$, which includes simple poles $(p_{1,2} = \pm j\omega_n)$ on the $j\omega$-axis. Hence, its impulse response persists in the form of oscillations at the natural frequency.

As another example of a marginally stable system, we consider a second-order system with a pole at the origin, described by: $G(s) = \frac{K}{s(\tau s + 1)}$. The impulse response of this system asymptotically approaches unity in the steady-state.

**Internal Stability.** The notion of internal stability is applicable to models of feedback control systems and requires that all signals within the closed-loop system remain bounded for all bounded inputs. This, in turn, implies that all relevant transfer functions between each individual input–output pair in the feedback control system are stable.

The internal stability requirements are met if the closed-loop characteristic polynomial (Chapter 4) is stable and there are no right-half plane pole-zero cancellations among the concatenated system transfer functions.

For general interconnected systems, the internal stability requires that the feedback loop has a net gain that is smaller than unity at all frequencies. In studies on robust control systems, this requirement has been formulated as the small gain theorem (see references).

**Stability of the Equilibrium Point.** For a general nonlinear system model, $\dot{x}(t) = f(x, u)$, stability refers to the stablity of equilibrium point defined by: $f(x_e, u_e) = 0$. In particular, the equilibium point is said to be stable if a system trajectory $x(t)$ that starts in the vicinity of $x_e$ stays close to $x_e$. The equilibrium point is said to be asymptotically stable if a system trajectory $x(t)$ that starts in the vicinity of $x_e$ converges to $x_e$.

Although linear methods of stability analysis may be extended to the nonlinear systems under the small signal approximation, in more rigorous terms, the nonlinear system stability is analyzed through the use of Lyapunov methods (see references).

## 2.3 Sinusoidal Response of a System

The sinusoidal response of a system refers to its response to a sinusoidal input of the form: $u(t) = \cos \omega_0 t$ or $u(t) = \sin \omega_0 t$.

To characterize the sinusoidal response of the system, let the system model be given as: $G(s)$. We assume a complex exponential input of the form: $u(t) = e^{j\omega_0 t}$ and $u(s) = \frac{1}{s - j\omega_0}$; then, the system output is given as: $y(s) = \frac{G(s)}{s - j\omega_0}$.

Let $p_i$, $i = 1, \ldots n$, denote the poles of the transfer function, $G(s)$; then, using PFE, the system response is given as: $y(s) = \frac{A_0}{s - j\omega} + \sum_{i=1}^{n} \frac{A_i}{s - p_i}$.

Following the inverses Laplace transform, the time response of the system is given as: $y(t) = A_0 e^{j\omega t} + \sum_{i=1}^{n} A_i e^{p_i t}$.

To proceed further, we assume that the system is BIBO stable, that is, $Re[p_i] < 0$, so that the transient response component dies out with time. Then, the steady-state response is given as:

$$y_{ss}(s) = \frac{A_0}{s - j\omega_0}, \quad y_{ss}(t) = A_0 e^{j\omega_0 t}, \quad \text{where } A_0 = G(j\omega_0).$$

**Frequency Response Function.** The frequency response function of a system with transfer function $G(s)$ is defined as: $G(j\omega) = G(s)|_{s=j\omega}$. The frequency response function is a complex function of a real frequency variable $\omega$; hence, it is described in the magnitude-phase form as: $G(j\omega) = |G(j\omega)|e^{\angle G(j\omega)}$.

Further, when computed at a given input frequency, $\omega = \omega_0$, the frequency response function, $G(j\omega_0)$, represents a complex number. Hence, the steady-state response to a complex sinusoid $u(t) = e^{j\omega_0 t}$ is given in terms of magnitude and phase as:

$$y_{ss}(t) = G(j\omega_0)e^{j\omega_0 t} = |G(j\omega_0)|e^{\angle G(j\omega)}e^{j\omega_0 t}$$

By Euler's identity, we have: $e^{j\omega_0 t} = \cos(\omega_0 t) + j\sin(\omega_0 t)$, that is, the sinusoidal input can be separated into two orthogonal components; therefore, the steady-state response to $u(t) = \cos \omega_0 t$ is given as:

$$y_{ss}(t) = |G(j\omega_0)| \cos(\omega_0 t + \angle G(j\omega_0))$$

Similarly, the steady-state response to $u(t) = \sin \omega_0 t$ is given as:

$$y_{ss}(t) = |G(j\omega_0)| \sin(\omega_0 t + \angle G(j\omega_0))$$

Thus, in the case of stable linear systems, the steady-state response to sinusoid of a certain frequency is a sinusoid at the same frequency scaled by the magnitude of the frequency response function and includes the phase contributed by the frequency response function.

### 2.3.1 Sinusoidal Response of Low-Order Systems

Using the frequency response function, $G(j\omega)$, the sinusoidal response of the first- and second-order systems is described as follows:

**First-Order System.** For a first-order system, let $G(s) = \frac{K}{\tau s+1}$; then, its frequency response function in the magnitude-phase form is given as:

$$G(j\omega) = \frac{K}{(j\omega\tau + 1)} = \frac{K}{|j\omega\tau + 1|} \angle - \tan^{-1} \omega\tau.$$

Hence, $|G(j\omega)| = \frac{K}{|1+j\omega\tau|}$ and $\angle G(j\omega) = -\tan^{-1}\omega\tau$

Let $u(t) = \sin \omega_0 t$; then, we have: $y_{ss}(t) = \frac{K}{\sqrt{\omega_0^2\tau^2+1}} \sin(\omega_0 t - \phi)$,

where $\phi = \tan^{-1}\omega_0\tau$.

**Second-Order System with a Pole at the Origin.** For a second-order system with a pole at the origin, let $G(s) = \frac{K}{s(\tau s+1)}$; then, its frequency response function in the magnitude-phase form is given as:

$$G(j\omega) = \frac{K}{j\omega (j\omega\tau + 1)} = \frac{K/\omega}{|j\omega\tau + 1|} \angle - 90° - \tan^{-1} \omega\tau$$

Hence, $|G(j\omega)| = \frac{K/\omega}{|1+j\omega\tau|}$ and $\angle G(j\omega) = -\tan^{-1}\omega\tau - 90°$

Let $u(t) = \sin \omega_0 t$; then, we have: $y_{ss}(t) = \frac{K/\omega_0}{\sqrt{\omega_0^2\tau^2+1}} \sin(\omega_0 t - \phi)$,

where $\phi = 90° + \tan^{-1}\omega_0\tau$.

**Second-Order System with Real Poles.** For a second-order system with real poles, let $G(s) = \frac{K}{(\tau_1 s+1)(\tau_2 s+1)}$; then, its frequency response function is given as:

$$G(j\omega) = \frac{K}{(j\omega\tau_1 + 1)(j\omega\tau_2 + 1)}$$

$$= \frac{K}{|j\omega\tau_1 + 1||j\omega\tau_2 + 1|} \angle - \tan^{-1} \omega\tau_1 - \tan^{-1} \omega\tau_2$$

Let $u(t) = \sin \omega_0 t$; then, we have: $y_{ss}(t) = \dfrac{K}{\sqrt{\omega_0^2 \tau_1^2 + 1}\sqrt{\omega_0^2 \tau_2^2 + 1}} \sin$

$(\omega_0 t - \phi)$, where $\phi = \tan^{-1} \omega_0 \tau_1 + \tan^{-1} \omega_0 \tau_2$.

**Second-Order System with Complex Poles.** For a second-order system with complex poles, let $G(s) = \dfrac{K}{(s+\sigma)^2 + \omega_d^2}$; then, its frequency response function is given as:

$$G(j\omega) = \frac{K}{(\sigma + j\omega)^2 + \omega_d^2}$$

$$= \frac{K}{|\sigma^2 + \omega_d^2 - \omega^2 + j2\sigma\omega|} \angle - \tan^{-1}\left(\frac{2\sigma\omega}{\sigma^2 + \omega_d^2 - \omega^2}\right)$$

Let $u(t) = \sin \omega_0 t$; then, we have: $y_{ss}(t) = \dfrac{K}{|\sigma^2 + \omega_d^2 - \omega_0^2 + j2\sigma\omega_0|} \sin$

$(\omega_0 t - \phi)$, where $\phi = \tan^{-1}\left(\dfrac{2\sigma\omega_0}{\sigma^2 + \omega_d^2 - \omega_0^2}\right)$

**Obtaining Sinusoidal Response in MATLAB.** The magnitude and phase of the transfer function at a given frequency can be obtained by using the "bode" command from the the MATLAB Control System Toolbox. The argument for the MALTAB "bode" command is a dynamic system object created by using the "tf" command, as illustrated by the following examples.

**Example 2.6:** A small DC motor.

We consider a small DC motor model with the following parameter values: $R = 1\Omega$, $L = 10$ mH, $J = 0.01$ kgm$^2$, $b = 0.1\frac{\text{Ns}}{\text{rad}}$, $k_t = k_b = 0.05$; then, $G(s) = \dfrac{500}{(s+10)(s+100)+25}$.

Further, the frequency response function for DC motor is given as: $G(j\omega) = \dfrac{0.488}{(0.01\omega+1)(0.097\omega+1)}$.

We consider a sinusoidal input: $u(t) = \sin 10t$; then, in the steady-state, we have:

$$y_{ss}(t) = 0.35 \sin(10t - 50°).$$

The magnitude and the phase can be obtained by using the following MATLAB script:

```
G=tf(500,[1 110 1025]);      % define transfer function
[mag,ph]=bode(G,10)          % obtain magnitude and phase
```

The sinusoidal response of the DC motor is plotted by using the following MATLAB script (Figure 2.6):

```
t=0:.01:10;                  % define time period
u=sin(10*t);                 % sinusoidal input signal
```

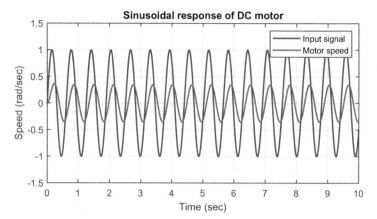

**Figure 2.6**    Sinusoidal response of DC motor.

```
y=lsim(G,u,t);                          % DC motor output
plot(t,u,t,y), grid                     % plot input and output
legend('Input signal','Motor speed')    % figure legend
```

**Example 2.7:** The mass–spring–damper system.

We consider a mass–spring–damper system, where the following parameter values are assumed: $m = 1$ kg, $b = 2\frac{Ns}{m}$, $k = 5\frac{N}{m}$. Then, the system transfer function is given as: $G(s) = \frac{1}{(s+1)^2+2^2}$, with complex poles located at: $s = -1 \pm j2$.

We consider a sinusoidal input: $u(t) = \sin \pi t$; then, in the steady-state, we have:

$$y_{ss}(t) = 0.126 \sin(\pi t - 128°).$$

The MATLAB script to obtain the magnitude and phase of the mass–spring–damper system sinusoidal response includes the following commands:

```
G=tf(1,[1 2 5]);        % define transfer function
[mag,ph]=bode(G,pi)     % obtain magnitude and phase
```

### 2.3.2 Visualizing the Frequency Response

We note that for a particular value of $\omega$, the frequency response, $G(j\omega)$, is a complex number, which is described in terms of its magnitude and phase as: $G(j\omega) = |G(j\omega)|e^{j\phi(\omega)}$.

Further, as $\omega$ varies from $0$ to $\infty$, both magnitude and phase may be plotted as functions of $\omega$ and are referred as the Bode magnitude and phase plots.

It is customary to plot both magnitude and phase against $\log \omega$, which accords a high dynamic range to $\omega$.

Further, the magnitude is plotted in decibels as: $|G(j\omega)|_{dB} = 20 \log_{10} |G(j\omega)|$. The phase $\phi(\omega)$ is plotted in degrees. The resulting frequency response plots are commonly known as Bode plots.

**Obtaining Bode Plot in MATLAB.** In the MATLAB Control Systems Toolbox, the Bode plot is obtained by using the "bode" command, invoked after defining the system transfer function, as follows:

```
G=tf(num,den);        % define system transfer function
bode(G),grid          % obtain Bode mag and phase plots
```

The plotting of the frequency response in magnitude and phase form is illustrated in the case of the first- and second-order systems below.

**First-Order System.** As a first-order transfer function, let $G(j\omega) = \frac{K}{(j\omega\tau+1)}$; then, the magnitude and phase responses are computed as:

$$|G(j\omega)|_{dB} = 20 \log_{10} K - 20 \log_{10} |1 + j\omega\tau|$$

$$\phi(\omega) = -\tan^{-1}(\omega\tau).$$

For $\omega \ll 1/\tau$, the $20 \log_{10} K$ is a constant value that defines the low-frequency asymptote on the Bode magnitude plot.

For $\omega \gg 1/\tau$, the $-20 \log_{10} |1+j\omega\tau|$ term has a slope of $-20$ dB/decade and defines the high-frequency asymptote on the Bode magnitude plot. These two asymptotes intersect at $\omega\tau = 1$, where $|G(j/\tau)|_{dB} = 20 \log K - 3$ dB.

The graph of $\phi(\omega)$ is characterized by the following points of interest:

$$\phi(0) = 0°, \ \phi(0.1/\tau) = -5.7°, \ \phi(1/\tau) = -45°,$$

$$\phi(10/\tau) = -84.3°, \ \phi(\infty) = -90°.$$

**Example 2.8:** A first-order system model.

Let $G(s) = \frac{1}{s+1}$; then, the Bode magnitude and phase plots (Figure 2.7(a)) are characterized by:

Low frequency: $G(j0) = 0$ dB $\angle 0°$
Corner frequency: $G(j1) = -3$ dB $\angle -45°$
High frequency: $G(j\infty) = -\infty$ dB $\angle -90°$

**Second-Order System with a Pole at the Origin.** For a second-order system with a pole at the origin, let $G(j\omega) = \frac{K}{j\omega(j\omega\tau+1)}$; then, the magnitude and

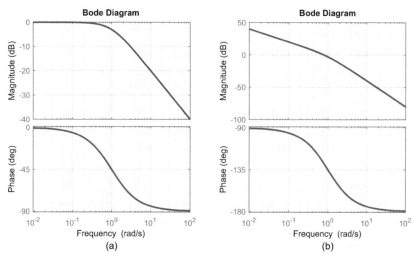

**Figure 2.7** The Bode magnitude and phase plots for: (a) a first-order system; (b) second-order system with a pole at the origin.

phase of the frequency response are computed as:

$$|G(j\omega)|_{dB} = 20\log_{10} K - 20\log_{10}\omega - 20\log_{10}|1 + j\omega\tau|$$
$$\phi(\omega) = -90° - \tan^{-1}(\omega\tau).$$

The slope of the Bode magnitude plot changes from $-20$ dB/decade to $-40$ dB/decade at the corner frequency, $\omega = 1/\tau$. The plot of $\phi(\omega)$ varies from $-90°$ to $-180°$, with $\phi(1/\tau) = -135°$.

**Example 2.9:** Second-order system with a pole at the origin.

Let $G(s) = \frac{1}{s(s+1)}$; then, the Bode magnitude and phase plots (Figure 2.7(b)) are characterized by:

Low frequency: $G(j0) = \infty$ dB $\angle 0°$
First corner frequency: $G(j1) = -3$ dB $\angle -135°$
High frequency: $G(j\infty) = -\infty$ dB $\angle -180°$

**Second-Order System with Real Poles.** For a second-order system with real poles, let $G(j\omega) = \frac{K}{(1+j\omega\tau_1)(1+j\omega\tau_2)}$; then, the magnitude and phase of the frequency response are computed as:

$$|G(j\omega)|_{dB} = 20\log_{10} K - 20\log_{10}|1 + j\omega\tau_1| - 20\log_{10}|1 + j\omega\tau_2|$$
$$\phi(\omega) = -\tan^{-1}(\omega\tau_1) - \tan^{-1}(\omega\tau_2).$$

The resulting frequency response plot is characterized by two corner frequencies: $\omega_1 = \frac{1}{\tau_1}, \omega_2 = \frac{1}{\tau_2}$. Further, the slope of the Bode magnitude

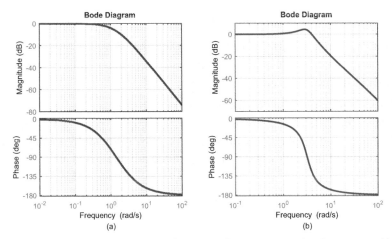

**Figure 2.8** The Bode magnitude and phase plots for: (a) a second-order system with real poles; (b) a second-order system with complex poles.

plot changes from 0 to $-20$ dB/decade to $-40$ dB/decade at the corner frequencies.

The plot of $\phi(\omega)$ varies from $0°$ to $-180°$. The phase at the corner frequencies is determined by their mutual proximity. For $\omega_1 \ll \omega_2$, we have: $\phi(\omega_1) = -45°$ and $\phi(\omega_2) = -135°$.

**Example 2.10:** Second-order system with real poles.

Let $G(s) = \frac{2}{(s+1)(s+2)}$; then, the Bode plot (Figure 2.8(a)) is characterized by:

Low frequency: $G(j0) = 0$ dB$\angle 0°$
Corner frequencies: $G(j1) = -10$ dB$\angle -72°$, $G(j1) = -16$ dB$\angle -108°$
High frequency: $G(j\infty) = -\infty$dB$\angle -180°$

**Second-Order System with Complex Poles.** For a second-order system with complex poles, let $G(s) = \frac{\omega_n^2}{s^2+2\zeta\omega_n s+\omega_n^2}$, where $\omega_n$ and $\zeta$ represent the natural frequency and the damping ratio; then, the magnitude and phase of the frequency response are computed as:

$$|G(j\omega)|_{\text{dB}} = -20\log_{10}\left|1 - \left(\frac{\omega}{\omega_n}\right)^2 + j2\zeta\left(\frac{\omega}{\omega_n}\right)\right|$$

$$\phi(\omega) = \tan^{-1}\left(\frac{2\zeta\left(\frac{\omega}{\omega_n}\right)}{\left[1 - \left(\frac{\omega}{\omega_n}\right)^2\right]}\right)$$

The resulting Bode plot (Figure 2.8(b)) is characterized by:

Low frequency: $G(j0) = 0$ dB$\angle 0°$
Corner frequencies: $G(j\omega_n) = -20\log_{10}|2\zeta|$ dB$\angle - 90°$
High frequency: $G(j\infty) = -\infty$dB$\angle - 180°$

**The Resonance Peak in the Frequency Response.** The Bode magnitude plot in the case of transfer function with complex poles with small enough damping ratio displays a distinctive peak at the resonant frequency, $\omega_r$, where the resonant frequency and peak magnitude are computed as:

$$\omega_r = \omega_n\sqrt{1 - 2\zeta^2}, \quad \zeta < \frac{1}{\sqrt{2}}$$

$$M_{p\omega} = \frac{1}{2\zeta\sqrt{1 - \zeta^2}}, \quad \zeta < \frac{1}{\sqrt{2}}.$$

**Example 2.11:** Second-order system with complex poles.

Let $G(s) = \frac{10}{s^2+2s+10}$; then, we have: $\omega_n = \sqrt{10}\frac{\text{rad}}{\text{sec}}$, $\zeta = \frac{1}{\sqrt{10}}$, $\omega_r = 8\frac{\text{rad}}{\text{sec}}$, and $M_{p\omega} = 1.67$ or $4.44$ dB (Figure 2.8(b)). The Bode magnitude and phase plots are characterized by:

Low frequency: $G(j0) = 0$ dB$\angle 0°$
Corner frequency: $G(j\sqrt{10}) = -6$ dB$\angle - 90°$
High frequency: $G(j\infty) = -\infty$dB$\angle - 180°$

The MATLAB script for the Bode plots for Examples 2.6–2.9 is given below:

```
G1=tf(1,[1 1]);              % first-order transfer function
G2=tf(1,[1 1 0]);            % 2nd order TF, pole at the origin
G3=tf(2,[1 3 2]);            % 2nd order TF, real poles
G4=tf(10,[1 2 10]);          % 2nd order TF, complex poles
subplot(1,2,1), bode(G1),grid % plot Bode plot
subplot(1,2,2), bode(G2),grid % plot Bode plot
subplot(1,2,1), bode(G3),grid % plot Bode plot
subplot(1,2,2), bode(G4),grid % plot Bode plot
```

**Relating Time and Frequency Responses.** When the system transfer function has a small damping ratio, the magnitude of the frequency response shows a resonant peak. Concurrently, the step response of the system becomes more oscillatory. This phenomenon is illustrated in the following example.

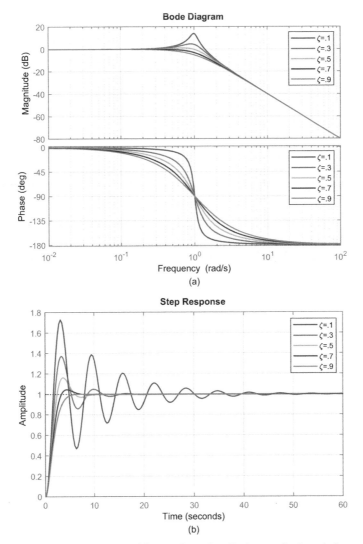

**Figure 2.9** Second-order system with complex poles: Bode magnitude and phase plots for selected damping ratios (a); accompanying unit-step response (b).

**Example 2.12:** An underdamped system.

Let $G(s) = \frac{1}{s^2 + 2\zeta s + 1}$, where $\zeta \in \{0.1, 0.3, 0.5, 0.7, 0.9\}$.

We compare the frequency response plots and the time-domain unit-step response for the second-order transfer functions with $\omega_n = 1$ for selected values of $\zeta \in \{0.1, 0.3, 0.5, 0.7, 0.9\}$ (Figure 2.9).

The time and frequency-domain plots were obtained by using the following MATLAB script:

```
figure(1), hold                % hold for multiple plots
figure(2), hold                % hold for multiple plots
for zi=.1:.2:.9,               % for loop
    G=tf(1,[1 2*zi 1]);        % vary damping
    figure(1), step(G),        % plot step response
    figure(2), bode(G),        % Bode plots
end                            % end for loop
grid                           % add grid and legend to the plot
legend('\zeta=.1','\zeta=.3','\zeta=.5','\zeta=.7','\zeta=.9')
```

From Figure 2.9, we observe that as the damping ratio drops below $\zeta = 0.5$:

(1) The peak on the Bode magnitude plot becomes prominent.
(2) The phase plot shows a sharper transition at corner frequency.
(3) The oscillations in the step response are more pronounced.

## Skill Assessment Questions

Link to the answers:
http://www.riverpublishers.com/book_details.php?book_id=449

1. Consider the first-order model of room heating with input $q_{in}$ and output $\Delta\theta$; assume the following parameter values: $C_r = 100[\frac{J}{\circ C}]$, $R_w = 0.2[\frac{\circ C}{W}]$;

   (a) Obtain an input-output transfer function for the system.
   (b) Solve and sketch the impulse response of the system.
   (c) Solve and sketch the step response of the system.
   (d) Compute steady-state system response to a sinusoidal input, $\sin \omega_0 t$, where $\omega_0 = 0.1$ rad/s.

2. Consider the model of a cylindrical hydraulic reservoir with a bottom flow valve with flow input $q_{in}$ and output $\Delta P$, where the following parameter values are assumed: $A = 1$ m$^2$, $\rho = 1[\frac{m^3}{s}]$, $R_l = 50[\frac{Ns}{m^5}]$; let $g \cong 10\frac{m}{s^2}$.

   (a) Obtain an input-output transfer function for the system.
   (b) Solve and sketch the impulse response of the system.
   (c) Solve and sketch the step response of the system.
   (d) Compute steady-state system response to a sinusoidal input, $\sin \omega_0 t$, where $\omega_0 = 1$ rad/s.

3. Consider the model of a small DC motor, where the following parameter values are assumed: $R = 1\Omega$, $L = 10$ mH, $J = 0.01$ kgm$^2$, $b = 0.1\frac{Ns}{rad}$, $k_t = k_b = 0.02$.

   (a) Obtain the transfer function for the DC motor from armature voltage to motor speed by ignoring armature inductance.
   (b) Solve and sketch the impulse response of the system.
   (c) Solve and sketch the step response of the system.
   (d) Compute steady-state system response to a sinusoidal input, $\sin \omega_0 t$, where $\omega_0 = 50$ rad/s.

4. Consider the model of a small DC motor, where the following parameter values are assumed: $R = 1\Omega$, $L = 10$ mH, $J = 0.01$ kgm$^2$, $b = 0.1\frac{Ns}{rad}$, $k_t = k_b = 0.02$.

   (a) Obtain the transfer function for the DC motor from armature voltage to motor speed without ignoring armature inductance.
   (b) Solve and sketch the impulse response of the system.

(c) Solve and sketch the step response of the system.

(d) Compute steady-state system response to a sinusoidal input, $\sin \omega_0 t$, where $\omega_0 = 50$ rad/s.

5. Consider the model of a spring-mass-damper system, where the following parameter values are assumed: $m = 1$ kg, $b = 2\frac{\text{Ns}}{\text{m}}$, $k = 2\frac{\text{N}}{\text{m}}$.

(a) Obtain the transfer function for the mass-spring-damper system with force input and position output.

(b) Solve and sketch the impulse response of the system.

(c) Solve and sketch the step response of the system.

(d) Compute steady-state system response to a sinusoidal input, $\sin \omega_0 t$, where $\omega_0 = 0.5$ rad/s.

6. Consider a series RLC circuit driven by a voltage source with capacitor voltage as output. Assume the following parameter values: $R = 2\Omega$, $L = 1$ mH, $C = 5$ mF

(a) Obtain the input–output transfer function for the circuit model.

(b) Solve and sketch the impulse response of the system.

(c) Solve and sketch the step response of the system.

(d) Compute steady-state system response to a sinusoidal input, $\sin \omega_0 t$, where $\omega_0 = \omega_r$, that is, the resonant frequency of the system.

7. Consider a parallel RLC circuit driven by a current source with capacitor voltage as output. Assume the following parameter values: $R = 2\Omega$, $L = 1$ mH, $C = 5$ mF

(a) Obtain the input–output transfer function for the circuit model.

(b) Solve and sketch the impulse response of the system.

(c) Solve and sketch the step response of the system.

(d) Compute steady-state system response to a sinusoidal input, $\sin \omega_0 t$, where $\omega_0 = \omega_r$, that is, the resonant frequency of the system.

8. Consider a notch filter employed to remove power line noise from an observed signal of interest, where the filter transfer function is given as:

$$G(s) = \frac{s^2 + 0.005\omega_n s + \omega_n^2}{s^2 + 0.5\omega_n s + \omega_n^2}, \omega_n = 120\pi.$$

(a) Solve and sketch the impulse response of the system.

(b) Solve and sketch the step response of the system.

(c) Compute steady-state filter response to a sinusoidal input: $u(t) = \sin 120\pi t$.

9. Consider the model of a car in cruise control with engine torque input and car speed output described by the transfer function: $G(s) = \frac{1/r}{ms+d}$, where $m$ is the mass of the car, $d$ is the slope of the drag curve vs. velocity, and $r$ is the radius of the wheel. Assume the following values: $m = 1500$ kg, $r = 0.3$ m, $d = 25$ kg/s.

(a) Solve and sketch the impulse response of the system.

(b) Solve and plot the car response to engine torque input of $500u(t)$.

(c) Compute steady-state car response to a sinusoidal input: $u(t) = \sin 0.1\pi t$ .

# 3

# Analysis of State Variable Models

## Learning Objectives

1. Formulate and analyze state variable models of dynamic systems.
2. Solve state equations in time-domain using the state-transition matrix.
3. Obtain the modal matrix and determine the stability of the model.
4. Obtain state-space realization of a transfer function model in alternate forms.

This chapter introduces the algebraic methods to analyze state variable models of dynamic systems. These models are described in terms of first-order ODEs involving time derivatives of a set of state variables (Chapter 1), that is, the natural variables associated with the energy storage elements or alternate variables used to describe the system.

In this chapter, we focus on the single-input single-output (SISO) systems that are described by a rational transfer function $G(s)$. However, the methods presented can be easily generalized to include multi-input multi-output (MIMO) systems (see references).

The state equations can be collectively integrated using a matrix exponential as an integrating factor. The matrix exponential defines the state-transition matrix of the system that contains the modes of system response and plays a fundamental role in the evolution of system trajectories. The solution to the homogeneous state equation includes a weighted sum of the columns of state transition matrix multiplied by initial conditions on state variables.

The state-transition matrix can be computed as the sum of an infinite series that is convergent for stable system models. The matrix can be alternatively computed using inverse Laplace transform of the resolvent matrix associated with the state variable description.

The solution to the state equations in the presence of an input is given as a convolutional integral that involves the state-transition matrix of the system.

The impulse response of the system can be similarly described in terms of its state-transition matrix.

Given a state variable model of the system, a transfer function model can be obtained by applying the Laplace transform. The degree of the denominator polynomial in the transfer function normally equals the number of state variables. However, in the unlikely event of a pole-zero cancellation, the transfer function may be of a lower order than the state variable model of the same physical system.

A given transfer function model can be realized into a state variable model in multiple ways, that is, the choice of state variables for a given transfer function model is not unique, and different choices of state variables accord different structures to the system matrix. The state variable vectors for these alternate models are related through bilinear transformations.

Specific realization structures may be preferred for ease of computing the system response or for determining its stability characteristics. For example, the modal realization reveals the natural modes of system response. The controller form realization facilitates the controller design using state variable methods. The diagonal representation decouples the system variables into a set of independent first-order ODE's that can be easily integrated.

This chapter describes the analysis techniques for state variable models of the continuous-time systems. These techniques are later extended to state variables models of sampled-data systems later in Chapter 7.

## 3.1 State Variable Models

The state equations of a system comprise first-order ODEs that describe dynamic system behavior in terms of time derivatives of a set of state variables. The number of state variables is selected as the minimum number of variables required to adequately describe the system behavior, which also equals the degree of its characteristic polynomial.

The natural variables associated with the energy storing system components are commonly selected as state variables, but alternate variables in equal number can be selected. Their time derivatives are expressed in terms of state and input variables. For linear time-invariant system models, the resulting state equations are commonly expressed in vector–matrix form.

Let $\mathbf{x}(t)$ describe a vector of state variables, $u(t)$ describe a scalar input, and $y(t)$ describe the output of a single-input single-output (SISO) system, then the state variable model of a linear time-invariant (LTI) dynamic system is written in its generic form as:

$$\dot{\mathbf{x}}(t) = \mathbf{A}\mathbf{x}(t) + \mathbf{b}u(t)$$
$$y(t) = \mathbf{c}^T\mathbf{x}(t) + du(t).$$

In the above, $\mathbf{A}$ is an $n \times n$ system matrix, where $n$ is the number of state variables, $\mathbf{b}$ is a column vector that distributes the inputs, $\mathbf{c}^T$ is a row vector ($T$ denotes the transpose operation) that combines the state variables into the output, and $d$ is a scalar feedforward term contributing to the output.

In a more general case, the state variable model of a multi-input multi-output (MIMO) system with $m$ inputs and $p$ outputs is described by state and output equations given as:

$$\dot{\mathbf{x}}(t) = \mathbf{A}\mathbf{x}(t) + \mathbf{B}\mathbf{u}(t)$$
$$\mathbf{y}(t) = \mathbf{C}\mathbf{x}(t) + \mathbf{D}\mathbf{u}(t)$$

In the above equations, the variable dimensions are given as: $x \in \mathbf{R}^n$, $u \in \mathbf{R}^m$, and $y \in \mathbf{R}^p$; the mappings defined by the matrices have the following dimensions: $\mathbf{A} \in \mathbf{R}^{n \times n}, \mathbf{B} \in \mathbf{R}^{n \times m}, \mathbf{C} \in \mathbf{R}^{p \times n}$, and $\mathbf{D} \in \mathbf{R}^{p \times m}$.

In the following, we restrict our discussion to models of SISO systems. Further, we assume that there is no direct contribution from input to output (i.e., $d = 0$). We note that it corresponds to a strictly proper transfer function for the ODE model.

### 3.1.1 Solution to the State Equations

We explore a time-domain solution to the state equations by integrating them from $0$ to $t$. In order to develop the desired time-domain solution, we may begin with a scalar differential equation:

$$\dot{x}(t) = ax(t) + bu(t).$$

Using an integrating factor $e^{-at}$, the equation is written as a total differential:

$$\frac{d}{dt}(e^{-at}x(t)) = e^{-at}bu(t)$$

Next, we assume an initial condition: $x(0) = x_0$, and integrate the equation from $\tau = 0$ to $\tau = t$ to obtain:

$$e^{-at}x(t) - x_0 = \int_0^t e^{-a\tau}bu(\tau)d\tau.$$

Hence, the solution to the scalar ODE involving a single variable, $x(t)$, is obtained as:

$$x(t) = e^{at}x_0 + \int_0^t e^{a(t-\tau)}bu(\tau)d\tau.$$

We note that the above ODE solution contains two parts: the first term on the right-hand side represents the system response to initial conditions and is termed as the zero-input solution. The second term is a convolution integral that describes the system response to input $u(t)$ and is termed as the zero-state solution.

Next, we explore the possibility to generalize this solution to the matrix case. For this purpose, we define a matrix exponential function as:

$$e^{\mathbf{A}t} = \sum_{i=0}^{\infty} \frac{\mathbf{A}^i t^i}{i!} = \mathbf{I} + \mathbf{A}t + \cdots$$

The infinite series defined above is convergent in the case of stable systems. Further, it can be verified that the matrix exponential obeys the matrix differential equation:

$$\frac{d}{dt}(e^{\mathbf{A}t}) = \mathbf{A}e^{\mathbf{A}t} = e^{\mathbf{A}t}\mathbf{A}.$$

Using the matrix exponential, the solution to the matrix equation, $\dot{\mathbf{x}}(t) = \mathbf{A}\mathbf{x}(t) + \mathbf{b}u(t)$, is written as:

$$\mathbf{x}(t) = e^{\mathbf{A}t}\mathbf{x}_0 + \int_0^t e^{\mathbf{A}(t-\tau)}\mathbf{b}u(\tau)d\tau.$$

The above time-domain solution has two parts: one that describes the system response to initial conditions $\mathbf{x}_0$, and the other that describes the system response to the input, $u(t)$.

In the absence of an input, that is, for $u(t) = 0$, the solution to the homogenous state equation: $\dot{\mathbf{x}}(t) = \mathbf{A}\mathbf{x}(t)$, and $\mathbf{x}(0) = \mathbf{x}_0$, is given as: $\mathbf{x}(t) = e^{\mathbf{A}t}\mathbf{x}_0$.

The matrix exponential $e^{\mathbf{A}t}$ that relates $\mathbf{x}(t)$ to $\mathbf{x}_0$ is called the state-transition matrix. Further, computation of the state-transition matrix constitutes an important first step toward writing the time-domain solution to the state equations.

## 3.1.2 Laplace Transform Solution and Transfer Function

Given the state variable model of a system, the Laplace transform can be used to solve the state equations as follows: by assuming an initial condition on the state variables, $x(0) = x_0$, and applying the Laplace transform to the state equations results in:

$$s\mathbf{x}(s) - \mathbf{x}_0 = \mathbf{A}\mathbf{x}(s) + \mathbf{b}u(s).$$

Next, the state variable vector is solved as:

$$\mathbf{x}(s) = (s\mathbf{I} - \mathbf{A})^{-1}\mathbf{x}_0 + (s\mathbf{I} - \mathbf{A})^{-1}\mathbf{b}u(s),$$

where $\mathbf{I}$ denote an $n \times n$ identity matrix.

The $n \times n$ matrix $(s\mathbf{I} - \mathbf{A})^{-1}$ appearing in the Laplace transform solution is called the resolvent matrix of $A$. Further, we note that the first term on the right hand side (rhs) involves the initial conditions on $x$, while the second term involves the input.

The application of the inverse Laplace transform then reveals the time-domain solution to the state equations that should match the matrix exponential solution obtained above. By comparing the two solutions, we arrive at the following relations:

$$\mathsf{L}[e^{\mathbf{A}t}] = (s\mathbf{I} - \mathbf{A})^{-1}$$

$$\mathsf{L}\left[\int_0^t e^{\mathbf{A}(t-\tau)}\mathbf{b}u(\tau)d\tau\right] = (s\mathbf{I} - \mathbf{A})^{-1}\mathbf{b}u(s).$$

The first equation above can be used to compute the state-transition matrix as:

$$e^{\mathbf{A}t} = \mathsf{L}^{-1}[(s\mathbf{I} - \mathbf{A})^{-1}].$$

Assuming zero initial conditions, the second equation is used to solve for the system input response as:

$$y(s) = \mathbf{c}^T(s\mathbf{I} - \mathbf{A})^{-1}\mathbf{b}u(s).$$

Since for a transfer function description, $y(s) = G(s)u(s)$ describes the input–output relation, by comparing the two, the system transfer function is obtained from its state variable description as:

$$G(s) = \mathbf{c}^T(s\mathbf{I} - \mathbf{A})^{-1}\mathbf{b}.$$

Using the inverse Laplace transform, the impulse response of the system, that is, its response to a unit-impulse, $\delta(t)$, is obtained as:

$$g(t) = \mathsf{L}^{-1}[G(s)] = \mathbf{c}^T e^{\mathbf{A}t}\mathbf{b}.$$

Further, in terms of the impulse response, the input response of the system is given by the convolution integral:

$$y(t) = \int_0^t g(t - \tau)u(\tau)d\tau.$$

In the more general case of a MIMO system, $G(s)$ represents a $p \times m$ transfer matrix, given as:

$$\mathbf{G}(s) = \mathbf{C}(s\mathbf{I} - \mathbf{A})^{-1}\mathbf{B} + \mathbf{D}.$$

By applying the inverse Laplace transform, the impulse response matrix is given as:

$$\mathbf{G}(t) = \mathbf{C}^T e^{\mathbf{A}t}\mathbf{B} + \mathbf{D}\delta(t).$$

Alternatively, using the Rosenbrock system matrix and the concept of transmission zeros (see references), the transfer matrix of a MIMO system can be obtained as:

$$\mathbf{G}(s) = \frac{\det \begin{bmatrix} s\mathbf{I} - \mathbf{A} & -\mathbf{B} \\ \mathbf{C} & \mathbf{D} \end{bmatrix}}{\det(s\mathbf{I} - \mathbf{A})}$$

Next, we discuss the properties of the state-transition matrix.

### 3.1.3 The State-Transition Matrix

The state-transition matrix, $e^{\mathbf{A}t}$, describes the time evolution of the state vector, $x(t)$. In order to explore the properties of the state-transition matrix, we may consider the homogenous state equation:

$$\dot{\mathbf{x}}(t) = \mathbf{A}\mathbf{x}(t), \quad \mathbf{x}(0) = \mathbf{x}_0.$$

The solution to the homogenous equation is given as: $\mathbf{x}(t) = e^{\mathbf{A}t}\mathbf{x}_0$.

The state-transition matrix of a linear time-invariant (LTI) system can be computed in the following ways:

1. $e^{\mathbf{A}t} = \mathsf{L}^{-1}[(s\mathbf{I} - \mathbf{A})^{-1}]$
2. $e^{\mathbf{A}t} = \sum_0^\infty \frac{\mathbf{A}^i t^i}{i!}$ (assuming series convergence)
3. By using the modal matrix (see Section 3.1.4)
4. By using the fundamental matrix (see References)
5. By using the Cayley–Hamilton theorem (see References)

Its Laplace transform, the resolvent of $\mathbf{A}$, can be computed in the following ways:

1. $(s\mathbf{I} - \mathbf{A})^{-1} = s^{-1}(\mathbf{I} + s^{-1}\mathbf{A} + \cdots)$

2. $(s\mathbf{I} - \mathbf{A})^{-1} = \frac{\text{adj}(s\mathbf{I}-\mathbf{A})}{\det(s\mathbf{I}-\mathbf{A})}$

**Characteristic Polynomial of A.** The characteristic polynomial of $\mathbf{A}$ is an $n$th order polynomial obtained as the determinant of the characteristic matrix $(s\mathbf{I} - \mathbf{A})$, that is,

$$\Delta(s) = |s\mathbf{I} - \mathbf{A}|$$

The roots of the characteristic polynomial are the eigenvalues of the system matrix $A$. Assuming no pole-zero cancellations, the characteristic polynomial matches the denominator polynomial in the transfer function, $G(s)$, where the latter is obtained as:

$$G(s) = \mathbf{c}^T \frac{adj\,(sI - A)}{|sI - A|}\mathbf{b} = \frac{n(s)}{d(s)}.$$

In the event of a pole-zero cancellation between the numerator and the denominator, the order of the denominator polynomial, $d(s)$, in the transfer function representation is less than $n$. Hence, the zeros of $d(s)$ form a subset of the eigenvalues of $A$.

**Obtaining State-Transition Matrix in MATLAB.** The state-transition matrix for a state variable description (A, B, C) may be obtained using symbolic algebra using the MATLAB Symbolic Math Toolbox. Assuming an A matrix has been defined, the state-transition matrix can be obtained in one of the following ways:

1. Using the matrix exponential, by using the following MATLB commands:

```
syms t            % declare symbolic variable t
eAt=expm(t*A)     % get state-transition matrix
```

2. Using the inverse Laplace transform, by using the following MATLAB commands:

```
syms s                  % define symbolic variable s
rA= inv(s*eye(n)-A);    % get resolvent of A
eAt=ilaplace(rA)        % get state transition matrix
```

**Obtaining System Transfer Function in MATLAB.** The system transfer function, $G(s)$, for the state variable description, $(A, B, C)$, where $A$ is an nxn matrix, can be obtained in the following ways:

1. Using the resolvent matrix by issuing the command:

   ```
   G(s)=C*inv(s*eye(n)-A)*B        % system transfer function
   ```

2. Using the Rosenbrock system matrix by issuing the following command:

```
G(s)= det([s*eye(n)-A -B;C 0])/det(s*eye(n)-A) % transfer function
```

### 3.1.4 Homogenous State Equation and Asymptotic Stability

The stability in the case of state variable model depends on the homogenous part of the state equation, which is described as:

$$\dot{x}(t) = Ax(t), \quad x(0) = x_0$$

We assume that the $A$ matrix has a full set of eigenvectors; these can be obtained by solving: $(\lambda_i I - A)v_i = 0$, $i = 1, \ldots, n$, where $\lambda_i$ denotes an eigenvalue of $A$.

The system natural response modes are described by $e^{\lambda_i t}$, $i = 1, \ldots, n$. Using arbitrary constants, $c_i$, the homogenous response of the system is described as:

$$x_h(t) = \sum_i^n c_i v_i e^{\lambda_i t}$$

**The Modal Matrix.** The modal matrix associated with a system of linear equations, $Ax = b$, is the $n \times n$ matrix whose columns are the eigenvectors of $A$. Accordingly, let $M = [v_1, v_2, \ldots, v_n]$ denote the modal matrix of eigenvectors of $A$ and define: $e^{\Lambda t} = diag([e^{\lambda_1 t}, e^{\lambda_2 t}, \ldots, e^{\lambda_n t}])$ as the diagonal matrix of the natural response modes; then, using an arbitrary constant vector $c$, the general solution to the homogenous system of equations is given as:

$$x_h(t) = Me^{\Lambda t}c$$

Assuming the known initial conditions: $x_h(0) = x_0 = Mc$; we have, $c = M^{-1}x_0$, that is, the particular solution to the homogenous system for the assumed initial conditions is given as:

$$x_h(t) = Me^{\Lambda t}M^{-1}x_0.$$

Therefore, the state-transition matrix can be computed from the modal matrix as:

$$e^{\mathbf{A}t} = \mathbf{M}e^{\mathbf{\Lambda}t}\mathbf{M}^{-1}.$$

**Asymptotic Stability.** The modal matrix can be used to describe the stability of the homogenous state variable equation described by: $\dot{\mathbf{x}}(t) = \mathbf{A}\mathbf{x}(t)$, $\mathbf{x}(0) = \mathbf{x}_0$.

The homogenous state equation is said to be asymptotically stable if $\lim_{t\to\infty} \mathbf{x}(t) = 0$. Further, since $\mathbf{x}(t) = e^{\mathbf{A}t}\mathbf{x}_0$, the homogenous system is asymptotically stable if $\lim_{t\to\infty} e^{\mathbf{A}t} = 0$.

Using the modal decomposition, $e^{\mathbf{A}t} = \mathbf{M}e^{\mathbf{\Lambda}t}\mathbf{M}^{-1}$, the homogenous system is asymptotically stable if $\lim_{t\to\infty} e^{\mathbf{\Lambda}t} = 0$. Further, since $e^{\mathbf{\Lambda}t} = diag([e^{\lambda_1 t}, e^{\lambda_2 t}, \ldots, e^{\lambda_n t}])$, the above condition implies that $Re[\lambda_i] < 0$, where $\lambda_i$ represents the roots of the characteristic polynomial:

$$\Delta(s) = |s\mathbf{I} - \mathbf{A}|.$$

Thus, the concept of asymptotic stability in the case of state variable models matches that of the BIBO stability in the case of transfer functions (Chapter 2).

**Modal Matrix in MATLAB.** The modal matrix containing eigenvectors of $\mathbf{A}$ is obtained by using the "eig" command in MATLAB. The eigenvectors so obtained are normalized to unity. Further, the state-transition matrix can be obtained from the modal matrix by using commands from the MATLAB Symbolic Math Toolbox as follows:

```
sympref('FloatingPointOutput',true);  % set output format
syms t real                           % declare symbolic variable t
[M,D]=eig(A);                         % modal matrix and eigenvalues
eAt=M*expm(t*D)/M                     % state-transition matrix
```

The computation of modal and state transition matrices is illustrated using the following examples.

**Example 3.1:** The mass–spring–damper system.

We consider the mass–spring–damper model (Example 1.8); the position of the mass, $x(t)$, and its velocity, $v(t) = \dot{x}(t)$ are selected as state variables. Then state variable model comprises the state and output equations as:

$$\frac{d}{dt}\begin{bmatrix} x \\ v \end{bmatrix} = \begin{bmatrix} 0 & 1 \\ -k/m & -b/m \end{bmatrix}\begin{bmatrix} x \\ v \end{bmatrix} + \begin{bmatrix} 0 \\ 1/m \end{bmatrix} f, \quad x = \begin{bmatrix} 1 & 0 \end{bmatrix}\begin{bmatrix} x \\ v \end{bmatrix}.$$

Then, characteristic matrix of the model is given as:

$$s\mathbf{I} - \mathbf{A} = \begin{bmatrix} s & -1 \\ \frac{k}{m} & s + \frac{b}{m} \end{bmatrix}.$$

The resolvent matrix of $\mathbf{A}$ is computed as:

$$(s\mathbf{I} - \mathbf{A})^{-1} = \frac{1}{\Delta(s)} \begin{bmatrix} s + \frac{b}{m} & 1 \\ \frac{-k}{m} & s \end{bmatrix};$$

$$\Delta(s) = s^2 + \frac{b}{m}s + \frac{k}{m}.$$

The computation of the modal and state-transition matrices is considered separately when the characteristic polynomial $\Delta(s)$ has real or complex roots.

**Polynomial with Real Roots.** We assume that: $m = 1$, $k = 2$, and $b = 3$; then, the characteristic polynomial is: $\Delta(s) = s^2 + 3s + 2$, which has real roots: $s_1, s_2 = -1, -2$.

The modal matrix of eigenvectors is given as: $\mathbf{M} = \begin{bmatrix} -1 & -1 \\ 1 & 2 \end{bmatrix}$. Let $\Lambda = \begin{bmatrix} -1 & 0 \\ 0 & -2 \end{bmatrix}$; then, $\mathbf{A} = \mathbf{M}\Lambda\mathbf{M}^{-1}$ and $e^{\mathbf{A}t} = \mathbf{M}e^{\Lambda t}\mathbf{M}^{-1}$. The resulting state-transition matrix is given as:

$$e^{\mathbf{A}t} = \begin{bmatrix} 2e^{-t} - e^{-2t} & e^{-t} - e^{-2t} \\ 2e^{-2t} - 2e^{-t} & 2e^{-2t} - e^{-t} \end{bmatrix}.$$

We note that the entries in the state transition matrix comprise the natural response modes: $\{e^{-t}, e^{-2t}\}$ of system response. Further, we can verify that: $\lim_{t \to \infty} e^{\mathbf{A}t} = 0$; hence, the homogenous state equation is asymptotically stable.

**Polynomial with Complex Roots.** Alternatively, let $m = 1$, $k = 2$, and $b = 2$; then, the characteristic polynomial is: $\Delta(s) = s^2 + 2s + 2$, which has complex roots: $s_1, s_2 = -1 \pm j1$. The complex eigenvectors are given as: $\begin{bmatrix} 1 \\ 1 \pm j1 \end{bmatrix}$. Indeed, let $\mathbf{V} = \begin{bmatrix} 1 & 1 \\ -1 + j & -1 - j \end{bmatrix}$; then $\mathbf{V}^{-1}\mathbf{A}\mathbf{V} = diag(s_1, s_2)$.

In this case, the complex algebra can be avoided by constructing a modal matrix from the real and imaginary parts of eigenvectors. Accordingly, let

$$\mathbf{M} = \begin{bmatrix} 1 & 0 \\ -1 & 1 \end{bmatrix}; \text{ define } \mathbf{M}^{-1}\mathbf{A}\mathbf{M} = \begin{bmatrix} -1 & 1 \\ -1 & -1 \end{bmatrix} = \boldsymbol{\Gamma}; \text{ then, } \mathbf{A} = \mathbf{M}\boldsymbol{\Gamma}\mathbf{M}^{-1}$$

and $e^{\mathbf{A}t} = \mathbf{M}e^{\boldsymbol{\Gamma}t}\mathbf{M}^{-1}$. The resulting state-transition matrix is given as:

$$e^{\mathbf{A}t} = \begin{bmatrix} e^{-t}(\cos t + \sin t) & e^{-t}\sin t \\ -2e^{-t}\sin t & e^{-t}(\cos t - \sin t) \end{bmatrix}.$$

The entries in the state-transition matrix similarly comprise the natural response modes: $\{e^{-t}\sin t, e^{-t}\cos t\}$. Further, we can verify that: $\lim_{t\to\infty} e^{\mathbf{A}t} = 0$; hence, the homogenous state equation is asymptotically stable.

The state-transition matrices for systems with real and complex eigenvalues can be obtained in the MATLAB Symbolic Math Toolbox by issuing the following commands:

```
syms t real                    % define symbolic variable t
A1=[0 1; -2 -3];               % define system matrix
eAt=expm(A1*t)                 % state transition matrix

A2=[0 1; -2 -2];               % define system matrix
[M,D]=eig(A2);                 % modal matrix and eigenvalues
eAt=M*expm(t*D)/M;             % state-transition matrix
eAt=simplify(eAt)              % simplify state-transition matrix
```

**Example 3.2:** The DC motor model.

We consider the small DC motor model, where the following component values are assumed: $R = 1\ \Omega$, $L = 10$ mH, $J = 0.01$ kgm$^2$, $b = 0.1\ \frac{\text{Ns}}{\text{rad}}$, $k_t = k_b = 0.05$.

The state variables for the DC motor are selected as the armature current, $i_a(t)$, and the motor angular velocity, $\omega(t)$. Then, the state and output equations for the DC motor are given as:

$$\frac{d}{dt}\begin{bmatrix} i_a \\ \omega \end{bmatrix} = \begin{bmatrix} -100 & -5 \\ 5 & -10 \end{bmatrix}\begin{bmatrix} i_a \\ \omega \end{bmatrix} + \begin{bmatrix} 100 \\ 0 \end{bmatrix}V_a, \omega = \begin{bmatrix} 0 & 1 \end{bmatrix}\begin{bmatrix} i_a \\ \omega \end{bmatrix}.$$

The computation of the resolvent matrix, the transfer function, and the state transition matrix proceeds as follows:

$$(s\mathbf{I} - \mathbf{A})^{-1} = \frac{1}{\Delta(s)}\begin{bmatrix} s+10 & -5 \\ 5 & s+100 \end{bmatrix};$$

$$\Delta(s) = s^2 + 110s + 1025$$

$$G(s) = \mathbf{c}^T(s\mathbf{I} - \mathbf{A})^{-1}\mathbf{b} = \frac{500}{s^2 + 110s + 1025}$$

$$e^{\mathbf{A}t} = \begin{bmatrix} 1.003 & 0.056 \\ -0.056 & -0.003 \end{bmatrix} e^{-99.72t} + \begin{bmatrix} -0.003 & -0.056 \\ 0.056 & 1.0003 \end{bmatrix} e^{-10.28t},$$

where $\{e^{-99.72t}, e^{-10.28t}\}$ represent the system's natural response modes. These modes correspond to the electrical and mechanical time constants of the DC motor: $\tau_e \cong 0.01$ s, $\tau_m \cong 0.1$ s.

The state-transition matrix and the transfer function above were obtained by using the following MATLAB commands:

```
syms s                      % declare symbolic variable s
syms t real                 % declare symbolic variable t
A=[-100 -5; 5 -10];         % define system matrix
[M,D]=eig(A);               % modal matrix and eigenvalues
eAt=M*expm(t*D)/M           % state-transition matrix
B=[100;0]; C=[0 1];         % define input and output matrices
G(s)=C*inv(s*eye(2)-A)*B    % system transfer function
```

### 3.1.5 System Response for State Variable Models

In this section, we consider the system response of a state variable model to initial conditions, as well as its response to impulse and step inputs.

Having obtained the state-transition matrix of the system, the initial condition response of the system is computed as:

$$\mathbf{x}(t) = e^{\mathbf{A}t}\mathbf{x}_0$$

i.e., the initial condition response is obtained from the homogenous state equation.

The input response of the system with zero initial conditions is obtained as:

$$y(s) = G(s)u(s) = \mathbf{c}^T(s\mathbf{I} - \mathbf{A})^{-1}\mathbf{b}u(s)$$

For an impulse input, we have: $u(t) = \delta(t), u(s) = 1$. Then, the system response in Laplace domain is given as:

$$y_{imp}(s) = \mathbf{c}^T(s\mathbf{I} - \mathbf{A})^{-1}\mathbf{b}$$

The unit-impulse response in time-domain is obtained as:

$$y_{imp}(t) = \mathbf{c}^T e^{\mathbf{A}t}\mathbf{b}$$

For a step input, we have: $u(t) = 1(t), u(s) = 1/s$. Then, the time derivate of the system response in Laplace domain is given as:

$$s x(s) = (s I - A)^{-1} b$$

The unit-step response in time-domain is obtained as:

$$y_{step}(t) = c^T \left( \int_0^t e^{A(t-\tau)} d\tau \right) b$$

Alternatively, the step response can be obtained by integration of the impulse response as:

$$y_{step}(t) = \int_0^t y_{imp}(\tau) d\tau$$

**Example 3.3:** The DC motor model.

We consider the small DC motor model in Example 3.2, where the following component values are assumed: $R = 1\ \Omega$, $L = 10$ mH, $J = 0.01$ kgm$^2$, $b = 0.1\ \frac{Ns}{rad}$, $k_t = k_b = 0.05$.

The state-transition matrix for the dc motor model is given as:

$$e^{At} = \begin{bmatrix} 1.003 & 0.056 \\ -0.056 & -0.003 \end{bmatrix} e^{-99.72t} + \begin{bmatrix} -0.003 & -0.056 \\ 0.056 & 1.0003 \end{bmatrix} e^{-10.28t},$$

We assume an initial condition: $x_0 = \begin{bmatrix} 0 \\ 1 \end{bmatrix}$ on the state variables; then, the system response is given as:

$$x(t) = e^{At} x_0 = \begin{bmatrix} 0.056 \left( e^{-99.7t} - e^{-10.3t} \right) \\ 1.003 e^{-10.3t} - 0.003 e^{-99.7t} \end{bmatrix}$$

The impulse response of the DC motor is computed as:

$$g(t) = c^T e^{At} b = 5.6 \left( e^{-99.7t} - e^{-10.3t} \right)$$

Next, let $V_a(t) = u(t)$; then, the step response of the DC motor is computed as:

$$y(t) = \int_0^t g(t - \tau) u(\tau) d\tau = 0.488 + 0.056 e^{-99.7t} - 0.544 e^{-10.3t}$$

The DC motor state variable response to initial conditions and its output response to impulse and step inputs are plotted below (Figures 3.1 and 3.2).

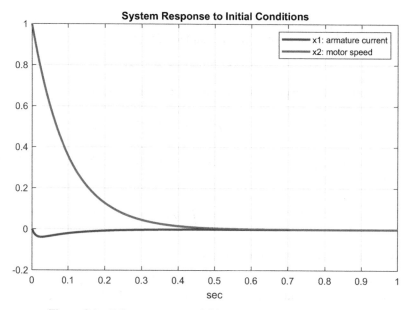

**Figure 3.1**    DC motor: state variable response to initial conditions.

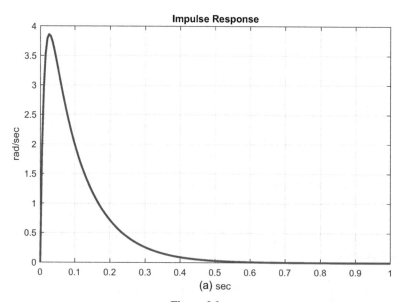

**Figure 3.2**

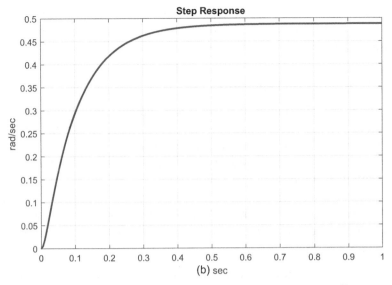

**Figure 3.2** DC motor: impulse response (a) and step response (b).

Assuming that the state variable model of the DC motor has been defined (Example 3.2), the initial condition response of the DC motor was obtained and plotted by using the following MATLAB commands:

```
t=0:.005:1;            % define time variable
x0=[0;1];              % define initial conditions
x=subs(eAt*x0);        % obtain state variables
plot(t',x)             % plot state variables
```

The impulse and step responses were plotted by using the following MATLAB commands:

```
yimp=C*subs(eAt*B);              % obtain impulse response
plot(t',subs(yimp)),grid         % plot impulse response
ystep=int(yimp,0,Inf)+int(yimp); % obtain step response
plot(t',subs(ystep)),grid        % plot step response
```

## 3.2 State Variable Realization of Transfer Function Models

In this section, we discuss the problem of realizing a transfer function model as a state variable model. Since the state variables for a physical system can be selected in multiple ways, the resulting state variable models have diverse model structures. Two useful structures, that is, the controller and the modal form structures, are discussed in detail in this section.

Obtaining a state variable description from a transfer function model is facilitated by first realizing the ODE model as a simulation diagram as discussed below.

## 3.2.1 Simulation Diagrams

A simulation diagram realizes an ODE model into a block diagram representation using static gains, integrators, summing nodes, and feedback loops. Simulation diagrams were historically used to simulate dynamic system models on analog computers that comprised integrators, summers, and potentiometers representing scalar gains.

There are multiple ways to realize a system model into a simulation diagram. All such diagrams are equivalent in that they represent the same input–output relationship described by the system transfer function. The two common realizations are described below.

**Serial Realization.** A serial chain of integrators can be used to realize the output of an ODE. To illustrate this idea, we consider a general second-order system model with complex poles described by its input–output transfer function:

$$\frac{y(s)}{u(s)} = \frac{K}{s^2 + 2\zeta\omega_n s + \omega_n^2} = \frac{n(s)}{d(s)}$$

where $n(s)$ and $d(s)$ represent the numerator and denominator polynomials.

The transfer function is converted into an ODE representation by cross-multiplying to obtain, $d(s)y(s) = n(s)u(s)$, followed by the application of inverse Laplace transform. The result is:

$$\ddot{y}(t) + 2\zeta\omega_n \dot{y}(t) + \omega_n^2 y(t) = Ku(t)$$

Next, we write an expression for the highest derivative in the equation as:

$$\ddot{y}(t) = -2\zeta\omega_n \dot{y}(t) - \omega_n^2 y(t) + Ku(t)$$

The resulting simulation diagram realizes the highest derivative using a summing node and uses a series of integrators to obtain the lower order derivatives and the output (Figure 3.3(a)).

**Parallel Realization.** A second-order transfer function, where the denominator polynomial has real roots: $s_{1,2} = -a_1, -a_2$, is as a transfer function:

$$\frac{y(s)}{u(s)} = \frac{K}{s^2 + (a_1 + a_2)s + a_1 a_2}$$

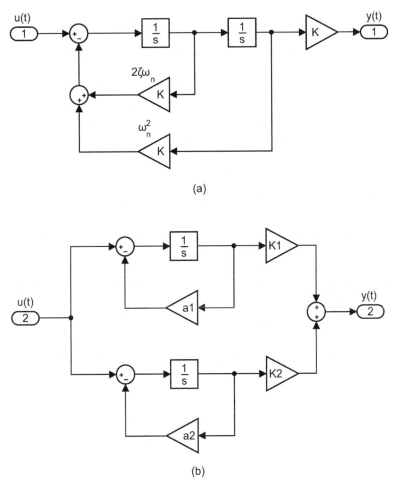

(a)

(b)

**Figure 3.3** Simulation diagram for realization of second-order transfer function: serial realization (a); parallel realization (b).

Using the partial fractions expansion (PFE), the above transfer function is expanded as:

$$\frac{y(s)}{u(s)} = \frac{K_1}{s + a_1} + \frac{K_2}{s + a_2}$$

In the case of parallel realization, the system output is expressed as: $y(s) = y_1(s) + y_2(s)$, where the two output components obey their respective ODEs: $a_i \dot{y}_i(t) + y_i(t) = K_i u(t), \quad i = 1, 2$, and share a common input, $u(t)$. The resulting parallel form realization is shown below (Figure 3.3(b)).

Having obtained the simulation diagram associated with an ODE model of a physical system, we designate the outputs of the integrators as state variables and their inputs as time derivatives of state variables, that is, the state equations.

The state equations resulting from the serial realization of a given transfer function acquire a structure known as the controller form; those resulting from the parallel realization acquire a modal structure. These structures are explained below in Sections 3.2.2–3.2.5.

### 3.2.2 Controller Form Realization

Having drawn a simulation diagram for the serial realization of a SISO system transfer function, a controller form state-space realization is obtained by selecting the outputs of the integrators as the state variables. The name controller form refers to the ease of designing a state feedback controller for the resulting state variable model (Chapter 10).

To illustrate the controller form realization, we consider a third-order transfer function:

$$G(s) = \frac{b_1 s^2 + b_2 s + b_3}{s^3 + a_1 s^2 + a_2 s + a_3}.$$

The above transfer function corresponds to the following ODE:

$$\dddot{y}(t) + a_1 \ddot{y}(t) + a_2 \dot{y}(t) + a_3 y(t) = b_1 \ddot{u}(t) + b_2 \dot{u}(t) + b_3 u(t)$$

Since the right-hand side includes derivatives of input variable, we use an auxiliary variable, $v(t)$, to split the ODE in two parts:

$$\dddot{v}(t) + a_1 \ddot{v}(t) + a_2 \dot{v}(t) + a_3 v(t) = u(t), \quad y(t) = b_1 \ddot{v}(t) + b_2 \dot{v}(t) + b_3 v(t)$$

The highest derivative term in the first ODE is realized using a summing node:

$$\dddot{v}(t) = -a_1 \ddot{v}(t) - a_2 \dot{v}(t) - a_3 v(t) + u(t)$$

The lower order derivative terms are realized using integrators. Once the simulation diagram for the first ODE is completed, the second ODE can be directly realized by summing the outputs of the integrators using gain elements (Figure 3.4).

Next, we designate the outputs of the integrators as state variables. Accordingly, let $x_1(t) = v(t), x_2(t) = \dot{v}(t), x_3(t) = \ddot{v}(t)$.

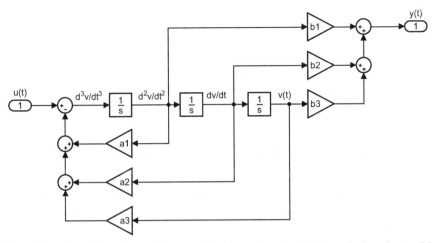

**Figure 3.4** Simulation diagram for controller form realization of the transfer function model.

Then, we obtain the following state equations from the diagram:

$$\dot{x}_1 = x_2$$
$$\dot{x}_2 = x_3$$
$$\dot{x}_3 = -a_3 x_1 - a_2 x_2 - a_3 x_3 + u$$

The output equation is given as:

$$y = b_1 x_1 + b_2 x_2 + b_3 x_3$$

The above equations are assembled in the vector–matrix form to describe the state variable model of the system:

$$\frac{d}{dt}\begin{bmatrix} x_1 \\ x_2 \\ x_3 \end{bmatrix} = \begin{bmatrix} 0 & 1 & 0 \\ 0 & 0 & 1 \\ -a_3 & -a_2 & -a_1 \end{bmatrix}\begin{bmatrix} x_1 \\ x_2 \\ x_3 \end{bmatrix} + \begin{bmatrix} 0 \\ 0 \\ 1 \end{bmatrix} u;$$

$$y = \begin{bmatrix} b_1 & b_2 & b_3 \end{bmatrix}\begin{bmatrix} x_1 \\ x_2 \\ x_3 \end{bmatrix}.$$

**General Controller Form Structure.** To generalize the above controller form realization to $n$ variables, we assume that the system transfer function is expressed as: $G(s) = \frac{n(s)}{d(s)}$, where $d(s) = s^n + a_1 s^{n-1} + \cdots + a_{n-1}s + a_n$. Then, by selecting the output variable and its derivatives: $y(t), \dot{y}(t), \ldots, y^{(n-1)}(t)$, as state variables, we obtain the following controller form structure for the state variable model of the system:

$$
\mathbf{A} = \begin{bmatrix} 0 & 1 & 0 & \cdots \\ 0 & 0 & 1 & \cdots \\ \vdots & \vdots & \ddots & 1 \\ -a_n & -a_{n-1} & \cdots & -a_1 \end{bmatrix}, \quad \mathbf{b} = \begin{bmatrix} 0 \\ 0 \\ \vdots \\ 1 \end{bmatrix}
$$

We make the following observations about the controller form structure:

1. In the controller form realization, the coefficients of the characteristic polynomial appear in reverse order in the last row of the system matrix.
2. The output equation contains the coefficients of the numerator polynomial.
3. The integrator outputs in the simulation diagram can be numbered from right to left (as above), or from left to right; the latter numbering reorders the state variables, so that the coefficients of the characteristic polynomial instead appear in the first row.
4. Given a transfer function model that is strictly proper, the associated controller form realization can be written by inspection.

**Controller Form Realization in MATLAB.** In the MATLAB Control Systems Toolbox, the controller form realization can be obtained by using the following commands. The state variables ordered from left to right, so that the coefficients of the denominator polynomial are placed in the first row of the system matrix:

```
G = tf(num, den);          % define plant transfer function
ss(G)                      % obtain controller form realization
```

**Example 3.4:** Mass–spring–damper system.

We consider the mass–spring–damper system, with the system transfer function given as: $G(s) = \frac{1}{s^2+2s+2}$.

Then, using the above MATLAB commands, we obtain a controller form state variable model, given as:

$$
\frac{d}{dt}\begin{bmatrix} x_1 \\ x_2 \end{bmatrix} = \begin{bmatrix} -2 & -2 \\ 1 & 0 \end{bmatrix} \begin{bmatrix} x_1 \\ x_2 \end{bmatrix} + \begin{bmatrix} 1 \\ 0 \end{bmatrix} u, \quad y = \begin{bmatrix} 0 & 1 \end{bmatrix} \begin{bmatrix} x_1 \\ x_2 \end{bmatrix}
$$

We note that the MATLAB supported realization numbers the integrators in the simulation diagram in reverse order. Hence, $x_1$ represents the velocity and $x_2$ the position of the inertial mass.

### 3.2.3 Dual (Observer Form) Realization

We assume that a state variable realization of a physical system model is given as:

$$\dot{\mathbf{x}}(t) = \mathbf{A}\mathbf{x}(t) + \mathbf{b}u(t), \quad y(t) = \mathbf{c}^T\mathbf{x}(t).$$

Then, its dual realization is defined by the following state variable model:

$$\dot{\mathbf{z}}(t) = \mathbf{A}^T\mathbf{z}(t) + \mathbf{c}u(t), \quad y(t) = \mathbf{b}^T\mathbf{z}(t).$$

We note that both the original and the dual models share the same scalar transfer function, that is:

$$G(s) = \mathbf{c}^T(s\mathbf{I} - \mathbf{A})^{-1}\mathbf{b} = \boldsymbol{b}^T(s\boldsymbol{I} - \boldsymbol{A}^T)^{-1}\boldsymbol{c} = G^T(s)$$

When the original state variable model is in the controller form, its dual appears as the "observer form realization," named so because of its use in the design of state observers.

**Observer Form Realization in MATLAB.** In the MATLAB Control Systems Toolbox, the observer form realization is termed as the companion form. The MATLAB command "canon" (for canonical) is used to transform a given model into its companion form, as given below:

```
G=tf(num, den);          % define plant transfer function
canon(G,'companion')     % transform into companion form
```

**Example 3.5:** Mass–spring–damper system.

We consider the mass–spring–damper system, with the system transfer function given as: $G(s) = \frac{1}{s^2+2s+2}$. Then, using the above MATLAB commands, we obtain an observer form state space model, given as:

$$\frac{d}{dt}\begin{bmatrix} x_1 \\ x_2 \end{bmatrix} = \begin{bmatrix} 0 & -2 \\ 1 & -2 \end{bmatrix}\begin{bmatrix} x_1 \\ x_2 \end{bmatrix} + \begin{bmatrix} 1 \\ 0 \end{bmatrix}u, \quad y = \begin{bmatrix} 0 & 1 \end{bmatrix}\begin{bmatrix} x_1 \\ x_2 \end{bmatrix}.$$

We may note that $x_1$ and $x_2$ in the above state variable model represent the dual variables, however, $x_2$ represents the position of the inertial mass.

### 3.2.4 Modal Realization

The modal realization represents a parallel structure for the system model composed of the first- and second-order factors. These factors, respectively, contain the real and complex eigenvalues of the system matrix. In order to develop the modal realization, the transfer function is first expanded into the first- and second-order factors using the PFE.

The second-order terms in the PFE contain complex roots, which may be realized in their controller form. Alternatively, the denominator of a second-order factor in the PFE is expressed as: $(s + \sigma)^2 + \omega^2$, which can be realized as a $2 \times 2$ block containing the real and imaginary parts of the eigenvalue as:

$$A_i = \begin{bmatrix} \sigma & \omega \\ -\omega & \sigma \end{bmatrix}.$$

The state variables representing individual factors are concatenated to obtain the modal realization for the system. The resulting modal matrix is block diagonal, that is, it consists of $1 \times 1$ and $2 \times 2$ blocks that correspond to the first- and second-order factors. The eigenvalues in these blocks represent the natural modes of system response. Further, they are represented in the system output, $y(t)$, through the output matrix.

The construction of the modal realization is illustrated in the following example.

**Example 3.6:** Modal realization.

To illustrate the modal realization, we consider a third-order system model with one real and two complex poles, which has the following partial fraction expansion:

$$G(s) = \frac{2(s^2 + s + 1)}{(s + 2)(s^2 + 2s + 2)} = \frac{3}{s + 2} - \frac{s + 2}{(s^2 + 2s + 2)}$$

Both factors can be realized individually in controller form and the results concatenated to obtain the following modal form realization:

$$\frac{d}{dt} \begin{bmatrix} x_1 \\ x_2 \\ x_3 \end{bmatrix} = \begin{bmatrix} -2 & 0 & 0 \\ 0 & 0 & 1 \\ 0 & -2 & -2 \end{bmatrix} \begin{bmatrix} x_1 \\ x_2 \\ x_3 \end{bmatrix} + \begin{bmatrix} 1 \\ 0 \\ 1 \end{bmatrix} u; \quad y = \begin{bmatrix} 3 & -2 & -1 \end{bmatrix} \begin{bmatrix} x_1 \\ x_2 \\ x_3 \end{bmatrix}$$

The $2 \times 2$ block corresponding to the quadratic factor in the PFE can be alternatively realized using real and imaginary parts of the complex eigenvalue. The resulting state variable model is given as:

$$\frac{d}{dt} \begin{bmatrix} x_1 \\ x_2 \\ x_3 \end{bmatrix} = \begin{bmatrix} -2 & 0 & 0 \\ 0 & -1 & 1 \\ 0 & -1 & -1 \end{bmatrix} \begin{bmatrix} x_1 \\ x_2 \\ x_3 \end{bmatrix} + \begin{bmatrix} 1 \\ 0 \\ 1 \end{bmatrix} u; y = \begin{bmatrix} 3 & -1 & -1 \end{bmatrix} \begin{bmatrix} x_1 \\ x_2 \\ x_3 \end{bmatrix}.$$

Further, we can verify that the two modal realizations share the same transfer function.

**Modal Realization in MATLAB.** The modal realization is obtained in the MATLAB Control Systems Toolbox by using the "canon" command, as illustrated below:

```
num=2*[1 1 1];             % define numerator polynomial
den= conv([1 2],[1 2 2]);  % define denominator polynomial
G=tf(num, den);            % define plant transfer function
canon(G,'modal')           % obtain modal form realization
```

## 3.2.5 Diagonalization and Decoupling

In the event when the denominator polynomial in the system transfer function has real and distinct roots, its modal matrix is a diagonal matrix with its eigenvalues placed on the main diagonal. The resulting state equations are described by a set of decoupled first-order ODEs that can be easily integrated, as illustrated by the following example.

**Example 3.7:** DC motor model.

We consider the transfer function of a small DC motor, given as: $G(s) = \frac{500}{s^2+10s+1025}$.

A PFE of the transfer function gives: $G(s) = 5.59 \left[\frac{1}{s+10.28} - \frac{1}{s+99.72}\right]$.
The modal realization is given as:

$$\begin{bmatrix} \dot{x}_1 \\ \dot{x}_2 \end{bmatrix} = \begin{bmatrix} -99.72 & 0 \\ 0 & -10.28 \end{bmatrix} \begin{bmatrix} x_1 \\ x_2 \end{bmatrix} + \begin{bmatrix} 1 \\ 1 \end{bmatrix} V_a, \omega = \begin{bmatrix} -5.59 & 5.59 \end{bmatrix} \begin{bmatrix} x_1 \\ x_2 \end{bmatrix}.$$

We note that the eigenvalues appearing on the main diagonal of the system matrix correspond to the DC motor electrical and mechanical time constants. The resulting DC motor model is described by first-order ODEs, given as:

$$\dot{x}_1(t) = -99.72x_1(t) + V_a$$
$$\dot{x}_2(t) = -10.28x_2(t) + V_a$$
$$\omega(t) = 5.59(x_2(t) - x_1(t))$$

We further note that the state variables in the diagonal realization are different than the armature current and motor speed originally used to model the system. In fact, the new state variables represent linear combinations of the original variables.

In the following, we discuss a general procedure to transform a given state variable model when we linearly combine the state variables to define a new set of variables.

## 3.3 Linear Transformation of State Variables

State variable models of dynamic systems can be transformed into equivalent models with different structures by changing the choice of variables. To study this phenomenon, we consider the general state variable model of a SISO system, described as:

$$\dot{\mathbf{x}}(t) = \mathbf{A}\mathbf{x}(t) + \mathbf{b}u(t)$$
$$y(t) = \mathbf{c}^T\mathbf{x}(t)$$

Linear transformation of the state variable vector $\mathbf{x}(t)$ involves multiplication of the state variable vector, $x(t)$, by a constant invertible matrix $\mathbf{P}$, resulting in a new set of state variables defined by $\mathbf{z}(t)$. Accordingly, let

$$\mathbf{z} = \mathbf{P}\mathbf{x}, \quad \mathbf{x} = \mathbf{P}^{-1}\mathbf{z}.$$

Using the above relations, the state and output equations are written in terms of the new state variable, $\mathbf{z}$, as:

$$\mathbf{P}^{-1}\dot{\mathbf{z}} = \mathbf{A}\mathbf{P}^{-1}\mathbf{z} + \mathbf{b}u, \quad y = \mathbf{c}^T\mathbf{P}^{-1}\mathbf{z}.$$

Multiplying on the left by $\mathbf{P}$ results in an alternate state variable model, given as:

$$\dot{\mathbf{z}} = \mathbf{P}\mathbf{A}\mathbf{P}^{-1}\mathbf{z} + \mathbf{P}\mathbf{b}u, y = \mathbf{c}^T\mathbf{P}^{-1}\mathbf{z}.$$

We note that the two models share the same transfer function, that is,

$$G(s) = \mathbf{c}^T\mathbf{P}^{-1}(s\mathbf{I} - \mathbf{P}\mathbf{A}\mathbf{P}^{-1})^{-1}\mathbf{P}\mathbf{b} = \mathbf{c}^T(s\mathbf{I} - \mathbf{A})^{-1}\mathbf{b}.$$

We also note that the transformed system matrix acquires an altered structure. This fact can be exploited to impart a desired structure to the system matrix through specifically defined linear transformation of the state variables.

To elaborate, let $\bar{\mathbf{A}} = \mathbf{P}\mathbf{A}\mathbf{P}^{-1}, \bar{\mathbf{b}} = \mathbf{P}\mathbf{b}$; the first equation is expressed as: $\bar{\mathbf{A}}\mathbf{P} = \mathbf{P}\mathbf{A}$. Given $\mathbf{A}, \mathbf{b}$, and the desired structure of $\bar{\mathbf{A}}, \bar{\mathbf{b}}$, the above relations may be used to define an appropriate transformation matrix, $\mathbf{P}$. In particular, the transformation to the controller and modal forms is explored below.

### 3.3.1 Transformation into Controller Form

A necessary condition to find a linear transformation matrix $\mathbf{P}$ to convert a given state variable model into controller form is that the system model is controllable, which is indicated if the following controllability matrix has full rank:

$$\mathbf{M}_C = [\mathbf{b}, \mathbf{Ab}, \ldots, \mathbf{A}^{n-1}\mathbf{b}].$$

We note that for the SISO systems, the controllability matrix is $n \times n$.

The controllability matrix is guaranteed to be of full rank if the transfer function description of the system has the same order, $n$, as the number of state variables used to model the system in state space.

To proceed further, we assume that the controllability matrix is of full rank, indicating that an equivalent controller form representation is attainable.

Let the controllability matrix for the desired controller form representation be given as: $\mathbf{M}_{CF}$. We note that $\mathbf{M}_{CF}$ contains the coefficients of the characteristic polynomial and can be written by inspection. Then, the matrix that transforms the given state-space model into its controller form representation is given as:

$$\mathbf{Q} = \mathbf{P}^{-1} = \mathbf{M}_C \mathbf{M}_{CF}^{-1}.$$

Transformation of a given state variable model into controller form is illustrated by the following example.

**Example 3.8:** DC motor model.

The state and output equations for a small DC motor model are given as:

$$\frac{d}{dt}\begin{bmatrix} i_a \\ \omega \end{bmatrix} = \begin{bmatrix} -100 & -5 \\ 5 & -10 \end{bmatrix}\begin{bmatrix} i_a \\ \omega \end{bmatrix} + \begin{bmatrix} 100 \\ 0 \end{bmatrix} V_a, \quad \omega = \begin{bmatrix} 0 & 1 \end{bmatrix}\begin{bmatrix} i_a \\ \omega \end{bmatrix}.$$

The controllability matrix for the model is given as: $\mathbf{M}_C = [\mathbf{b}, \mathbf{Ab}] = \begin{bmatrix} 100 & -10^4 \\ 0 & 500 \end{bmatrix}$.

We can verify that the controllability matrix is of full rank; hence, the model is controllable. Next, from the transfer function description of the DC motor, $G(s) = \frac{500}{s^2+110s+1025}$, we obtain the controller form realization as:

$$\dot{\mathbf{x}} = \begin{bmatrix} 0 & 1 \\ -1025 & -110 \end{bmatrix}\mathbf{x} + \begin{bmatrix} 0 \\ 1 \end{bmatrix} V_a, \quad \omega = \begin{bmatrix} 500 & 0 \end{bmatrix}\mathbf{x}.$$

The controllability matrix for the controller form realization is given as:

$$\mathbf{M}_{CF} = \begin{bmatrix} 0 & 1 \\ 1 & -110 \end{bmatrix}.$$

Then, the state transformation matrix for the DC motor model is computed as:

$$\mathbf{Q} = \mathbf{P}^{-1} = \begin{bmatrix} 1000 & 100 \\ 500 & 0 \end{bmatrix}, \quad \mathbf{P} = \begin{bmatrix} 0 & 0.002 \\ 0.01 & -0.02 \end{bmatrix}.$$

Indeed, $\overline{\mathbf{A}} = \mathbf{PAQ}$ matches the system matrix in the controller form.

### 3.3.2 Transformation into Modal Form

A matrix that has a full set of eigenvectors is diagonalizable by a linear transformation matrix $\mathbf{P}$ when the eigenvectors of $\mathbf{A}$ are selected as the columns of $\mathbf{P}^{-1}$.

The resulting state variable description, given as: $\overline{\mathbf{A}} = \mathbf{PAP}^{-1}, \overline{\mathbf{b}} = \mathbf{Pb}$, has a diagonal structure.

In the event when the $\mathbf{A}$ matrix has complex eigenvalues, its eigenvectors are also complex. However, by placing the real and imaginary parts of the eigenvector into columns of $\mathbf{P}^{-1}$, the matrix is transformed into modal form with real entries, where the corresponding $2 \times 2$ block contains the real and imaginary parts of the complex eigenvalues. This is illustrated using the following example.

**Example 3.9:** The mass–spring–damper system.

We consider the mass–spring–damper system with mass position $x(t)$ and velocity $v(t)$ selected as state variables. The state and output equations of the system for a particular choice of system parameters are given as:

$$\frac{d}{dt} \begin{bmatrix} x \\ v \end{bmatrix} = \begin{bmatrix} 0 & 1 \\ -2 & -2 \end{bmatrix} \begin{bmatrix} x \\ v \end{bmatrix} + \begin{bmatrix} 0 \\ 1 \end{bmatrix} u, \quad y = \begin{bmatrix} 1 & 0 \end{bmatrix} \begin{bmatrix} x \\ v \end{bmatrix}.$$

The eigenvalues of the system matrix are: $-1 \pm j1$; the complex eigenvectors are: $\begin{bmatrix} 1 \\ 1 \pm j1 \end{bmatrix}$.

By choosing $\mathbf{P}^{-1} = \begin{bmatrix} 1 & 0 \\ -1 & 1 \end{bmatrix}$, the modal form system matrix is obtained as: $\mathbf{PAP}^{-1} = \begin{bmatrix} -1 & 1 \\ -1 & -1 \end{bmatrix}$.

**Example 3.10:** DC motor model.

We consider the state variable model of a small DC motor (Example 3.2), given as:

$$\frac{d}{dt}\begin{bmatrix} i_a \\ \omega \end{bmatrix} = \begin{bmatrix} -100 & -5 \\ 5 & -10 \end{bmatrix}\begin{bmatrix} i_a \\ \omega \end{bmatrix} + \begin{bmatrix} 100 \\ 0 \end{bmatrix}V_a, \omega = \begin{bmatrix} 0 & 1 \end{bmatrix}\begin{bmatrix} i_a \\ \omega \end{bmatrix}.$$

Next, we use the MATLAB "eig" command to obtain the $P$ matrix to diagonalize the state variable model as:

```
A=[-100 -5; 5 -10];        % define system matrix
[V,D]=eig(A)               % diagonalize system matrix
```

The matrix $V = P^{-1} = \begin{bmatrix} -0.9985 & 0.0556 \\ 0.0556 & -0.9985 \end{bmatrix}$ obtained from MATLAB is then used to transform the system matrix $A$, and the input $b$ and output $c^T$ vectors. The result is a decoupled state variable description, given as:

$$\begin{bmatrix} \dot{x}_1 \\ \dot{x}_2 \end{bmatrix} = \begin{bmatrix} -99.72 & 0 \\ 0 & -10.28 \end{bmatrix}\begin{bmatrix} x_1 \\ x_2 \end{bmatrix} + \begin{bmatrix} -100.47 \\ -5.6 \end{bmatrix}V_a,$$

$$\omega = \begin{bmatrix} 0.056 & -0.998 \end{bmatrix}\begin{bmatrix} x_1 \\ x_2 \end{bmatrix}.$$

The decoupled system of equations includes two scalar ODEs that can be easily integrated. Assuming, for example, a unit-step input, the solution to the state variables is given as:

$$x_1(t) = 1.0075 + (x_{10} - 1.0075)e^{-t/99.72}$$
$$x_2(t) = 0.545 + (x_{20} - 0.545)e^{-t/10.28}.$$

In the above, $x_{10}, x_{20}$ represent the initial conditions on the state variables.

The system output represents a weighted sum of state variables and is computed as:

$$\omega(t) = 0.056x_1(t) - 0.998x_2(t).$$

The original DC motor state variables are recovered as: $\begin{bmatrix} i_a \\ \omega \end{bmatrix} =$ $P^{-1}\begin{bmatrix} x_1 \\ x_2 \end{bmatrix}$, that is,

$$i_a(t) = -0.9985x_1(t) + 0.0556x_2(t)$$
$$\omega(t) = 0.0556x_1(t) - 0.9985x_2(t).$$

## Skill Assessment Questions

Link to the answers:
http://www.riverpublishers.com/book_details.php?book_id=449

1. The state variable model of a simple pendulum is given as:

$$\frac{d}{dt}\begin{bmatrix} \theta \\ \omega \end{bmatrix} = \begin{bmatrix} 0 & 1 \\ -\frac{g}{l} & 0 \end{bmatrix}\begin{bmatrix} \theta \\ \omega \end{bmatrix} + \begin{bmatrix} 0 \\ 1 \end{bmatrix} T$$

Assume the following parameter values: $m = 1, g = 10$.

(a) Find the characteristic polynomial and the modes of system's natural response.

(b) Find the modal and state-transition matrices for the model.

(c) Transform the state variable model into modal form.

2. The state equation for the simplified model of a DC motor is given below.

$$\frac{d}{dt}\begin{bmatrix} \theta \\ \omega \end{bmatrix} = \begin{bmatrix} 0 & 1 \\ 0 & -\frac{1}{\tau_m} \end{bmatrix}\begin{bmatrix} \theta \\ \omega \end{bmatrix} + \begin{bmatrix} 0 \\ \frac{k_t}{JR} \end{bmatrix} V_a$$

Assume the following parameter values: $R = 1, J = 0.01, b = 0.1, k_t = k_b = 0.02$.

(a) Find the characteristic polynomial and the modes of system natural response.

(b) Find the state-transition matrix for the model.

(c) Compute and plot the state variables in response to a step input.

3. A series RLC circuit is described by the following equations, where $i_L, v_C$ represent the inductor current and the capacitor voltage:

$$L\frac{di(t)}{dt} + Ri(t) + V_c(t) = V_s(t), C\frac{dV_c}{dt} = i(t)$$

Assume the following parameter values: $R = 1\Omega, L = 1\text{mH}, C = 1\mu\text{F}$

(a) Develop a state variable model of the circuit.

(b) Find the modes of system natural response.

(c) Find the modal and state-transition matrices for the model.

4. A band pass RLC network is described by the following equations, where $i_L, v_C$ represent the inductor current and the capacitor voltage:

$$C\frac{dv_C}{dt} = \frac{V_s - v_C}{R} - i_L, \quad L\frac{di_L}{dt} = v_C$$

Assume the following parameter values: $R = 1\Omega, L = 1\text{mH}, C = 1\mu\text{F}$

(a) Develop a state variable model of the circuit.

(b) Find the modes of system natural response.

(c) Find the modal and state-transition matrices for the model.

5. The approximate model of an industrial process with delay is obtained as: $G(s) = \frac{20(1-0.5s)}{(0.5s+1)^2}$.

(a) Find the modes of system natural response.

(b) Draw simulation diagrams for serial and parallel realizations of the model.

(c) Obtain controller and modal form realizations for the model.

6. The postural dynamics in the sagittal plane are modeled as a rigid inverted pendulum, given as: $G(s) = \frac{k}{s^2-\Omega^2}$, where $k = 0.01$, $\Omega = 3 \text{ rad/sec}$.

(a) Find the modes of system natural response.

(b) Draw simulation diagrams for serial and parallel realizations of the model.

(c) Obtain controller and modal form realizations for the model.

7. The model of a flexible structure is given as: $G(s) = \frac{0.5}{s} + \frac{0.1}{s^2+s+100}$

(a) Find the modes of system natural response.

(b) Draw a simulation diagram for the parallel realization of the model.

(c) Obtain the modal form realization for the model.

8. Consider the following transfer function model: $G(s) = \frac{s+1}{s(s+2)(s+5)}$.

(a) Find the modes of system natural response.

(b) Draw simulation diagrams for serial and parallel realizations of the model.

(c) Obtain controller and modal form realizations for the model.

9. Consider the following transfer function model: $G(s) = \frac{s+1}{s(s^2+2s+2)}$.

(a) Find the modes of system natural response.

(b) Draw simulation diagrams for serial and parallel realization of the model.

(c) Obtain controller and modal form realizations for the model.

10. Consider the following transfer function model: $G(s) = \frac{28s+120}{s^2+7s+14}$.

(a) Find the modes of system natural response.

(b) Draw simulation diagrams for serial and parallel realization of the model.

(c) Obtain controller and modal form realizations for the model.

# 4

---

# Feedback Control Systems

---

## Learning Objectives

1. Characterize feedback controller models for single-input single-output (SISO) systems.
2. Characterize the PID controller in terms of the three basic control modes.
3. Characterize rate feedback controller for a SISO.

A control system is designed to monitor and regulate the response of a physical system, such as an automobile or an airplane. Automatic control systems invariably involve feeding back the output to compare with a reference input that the output is expected to follow. A comparator generates an error signal that is expected to go to zero in stead-state.

The traditional design of a single-input single-output (SISO) control system involves a controller, $K(s)$, in cascade with the plant, $G(s)$, in the feedback loop (Figure 4.1). The controller suitably conditions the error signal, $e = r - Hy$, obtained from a comparator that monitors and compares the measured output, $H\ y(t)$, where $H$ represents the sensor gain, against the reference input, $r(t)$.

The elementary controller for the SISO system is a static gain controller, $K$, that suitably affects the closed-loop characteristic polynomial and thereby the output of the system (Figure 4.1(a)). The static gain controller is simple to design and is effective in diverse control problems, such as regulation and reference tracking problems. If, however, the static gain controller does not meet the control objectives, then instead a dynamic controller, $K(s)$, may be considered.

A dynamic controller, described by the controller transfer function, $K(s)$, represents a dynamic system in its own right. The input to the dynamic controller is the error signal, $e(t)$, and its output, $u(t)$, is the input to the plant. A proportional-integral-derivative (PID) controller that comprises three basic modes of control is an example of the dynamic controller. Further, a dynamic controller is similar to a frequency selective filter.

93

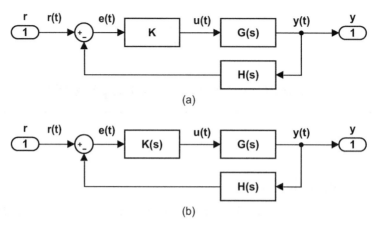

**Figure 4.1**    Feedback control system with static gain controller (a); with dynamic controller $K(s)$, plant $G(s)$, and sensor $H(s)$ (b).

The output feedback controller design has limited potential to affect the behavior of the closed-loop system, which it does through modifying the closed-loop characteristic polynomial. Pole placement using full state feedback is a more powerful controller design method that is employed with state variable models and allows arbitrary placement of closed-loop poles of the system (more on this topic in Chapter 9).

Rate sensors, such as rate gyros and tachometers, are commonly employed in position control applications. A feedback controller with rate feedback offers additional design flexibility in terms of transient as well as steady-state response improvement.

The cascade and rate controllers for transfer function models can be designed by using root locus technique or frequency response design methods. These are discussed in later in Chapters 6 and 9, respectively. The advent of digital computers in recent years has popularized alternate time-domain design methods that employ state variable models of control systems (more on this topic in Chapter 7).

The dynamic controllers designed using conventional methods are traditionally implemented using frequency selective filters built with operational amplifier circuits and RC networks. Nowadays, however, both filters and controllers are implemented using microcontrollers and industrial computers. The time-domain controller design methods also facilitate computer implementation of the controller (more on this topic in Chapter 8).

In this chapter, we will limit our discussion to the cascade and rate feedback controllers of the static and dynamic types. These controller are described below.

## 4.1 Static Gain Controller

A static gain controller consists of placing an amplifier with static gain, $K$, in cascade with the plant (Figure 4.1(a)) that affects the loop gain and hence the coefficients of the characteristic polynomial that determines the closed-loop system response. The gain $K$ can be selected keeping in view the desired and/or achievable root locations of the closed-loop characteristic polynomial.

The advantage of a static controller is its simplicity and ease of use in designing control systems. Its disadvantage is its limited ability to affect the behavior, that is, it offers a limited choice of achievable closed-loop pole locations.

Nevertheless, static controller is the first choice of the control systems designer in an effort to suitably modify the plant behavior to meet the transient response and/or steady-state error specifications.

To assess the effectiveness of the static gain controller, let $G(s) = \frac{n(s)}{d(s)}$ describe the plant transfer function, and let $K(s) = K$ describe the controller. Then, assuming $H(s) = 1$, the closed-loop transfer function is given as:

$$\frac{y(s)}{r(s)} = \frac{KG(s)}{1 + KG(s)} = \frac{Kn(s)}{d(s) + Kn(s)}.$$

The resulting closed-loop characteristic polynomial is given as: $\Delta(s, K) = d(s) + Kn(s)$.

Further, the error transfer function from the reference input to the comparator output is given as:

$$\frac{e(s)}{r(s)} = \frac{d(s)}{d(s) + Kn(s)}.$$

From the above relations, we note that:

1. The closed-loop poles are the roots of the characteristic polynomial, given as $\Delta(s, K) = d(s) + Kn(s)$, and are affected by the choice of $K$.
2. The static controller does not affect the zeros of the plant, that is, the roots of the numerator polynomial. In particular, any unstable zeros are not affected by the static controller.

3. It is generally advantageous to choose a large $K$ to increase the loop gain in order to reduce the steady-state tracking error, but this may have undesirable consequences. In particular, a high loop gain reduces the dynamic stability, that is, it makes the system response prone to oscillations.

**Example 4.1:** Environmental control system.

We assume that a model for an environmental control system is given as: $G(s) = \frac{1}{20s+1}$.

Then, using a static gain controller in the feedback loop, the characteristic polynomial the closed-loop system is obtained as: $\Delta(s, K) = 20s + 1 + K$.

Let a desired characteristic polynomial be given as: $\Delta_{des}(s) = 20(s + 5)$. Then, by comparing the coefficients, we obtain the static controller as: $K = 99$.

**Example 4.2:** Position control system.

We consider the model for a position control system given as: $G(s) = \frac{1}{s(0.1s+1)}$.

Then, using a static gain controller in the feedback loop, the characteristic polynomial of the closed-loop system is obtained as: $\Delta(s, K) = s(0.1s + 1) + K$.

We can equivalently express the characteristic polynomial as: $\Delta(s, K) = s^2 + 10s + 10K$.

Let a desired characteristic polynomial be given as: $\Delta_{des}(s) = s^2 + 10s + 50$.

Then, by comparing the coefficients, we obtain the static controller as: $K = 5$.

## 4.2 Dynamic Controllers

A dynamic controller is a dynamic system described by a transfer function, $K(s)$, that suitably conditions the error signal to generate the plant input signal, thus affecting the closed-loop system response.

To proceed further, we assume that the plant is described by its transfer function, $G(s) = \frac{n(s)}{n(s)}$, and the dynamic controller is described by its transfer function, $K(s) = \frac{n_c(s)}{d_c(s)}$; then, the closed-loop characteristic polynomial is given as:

$$\Delta(s) = d(s)d_c(s) + n(s)n_c(s).$$

The closed-loop transfer function from the reference input to the plant output is given as:

$$\frac{y(s)}{r(s)} = \frac{n(s)n_c(s)}{d(s)d_c(s) + n(s)n_c(s)}, \quad \Delta(s) = d(s)d_c(s) + n(s)n_c(s).$$

The error transfer function from the reference input to the comparator output is given as:

$$\frac{e(s)}{r(s)} = \frac{d(s)d_c(s)}{d(s)d_c(s) + n(s)n_c(s)}.$$

From the above relations, we observe that:

1. The closed-loop characteristic polynomial, $\Delta(s) = d(s)d_c(s) + n(s)$ $n_c(s)$, allows greater flexibility in the choice of closed-loop poles compared to the static gain controller.
2. The closed-loop input–output transfer function includes controller zeroes. This fact may be used to choose controller zeros at undesirable but stable plant pole locations to exclude them from the loop transfer function. However, such pole-zero cancellations are only good on paper and should absolutely be avoided in the case of unstable plant poles.

In the following, we discuss the characteristics of some common structures used to design dynamic controllers.

### 4.2.1 First-Order Phase-Lead and Phase-Lag Controllers

Traditionally, first-order dynamic controllers are categorized as phase-lead or phase-lag type. The lead and lag terms refer to the phase contribution from the controller to the Bode phase plot of the loop transfer function, defined as: $L(s) = KGH(s) = K(s)G(s)H(s)$. The phase contribution is positive for phase-lead controller and negative for phase-lag controller (Figure 4.2).

The positive phase from the phase-lead controller increases the phase margin and adds to dynamic stability. The negative phase from the phase-lag controller compromises stability, and is kept to a minimum while increasing the system error constants.

Both phase-lead and phase-lag controllers are first-order controllers, that is, they add a pole–zero pair to the loop transfer function, and can be described by a generic transfer function description:

$$K(s) = \frac{K(s + z_c)}{s + p_c},$$

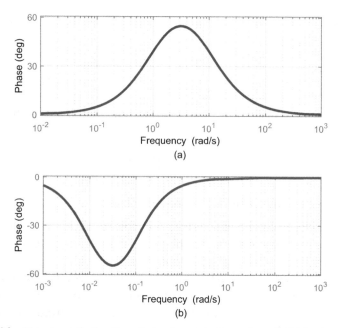

**Figure 4.2** Phase contribution from first-order controllers; phase-lead (a); phase-lag (b).

where the phase-lead controller is characterized by $z_c < p_c$, and the phase-lag controller is characterized by $z_c > p_c$. Further, the phase contribution of the controller is given as $\angle K = \theta_z - \theta_p$, where $\theta_z$ and $\theta_p$ denote the angles subtended by the controller pole and zero at the desired closed-loop pole location.

As an example, a dynamic controller given as $K(s) = \frac{s+1}{s+10}$ is of phase-lead type, whereas $K(s) = \frac{s+0.1}{s+0.01}$ is of phase-lag type. The phase contribution of the two controllers is shown below (Figure 4.2).

A lead-lag controller is a second-order controller that combines a phase-lead section with a phase-lag section and thus has the form:

$$K(s) = \frac{K(s+z_1)(s+z_2)}{(s+p_1)(s+p_2)}.$$

For example, a lead-lag controller is given as: $K(s) = \frac{K(s+0.1)(s+10)}{(s+1)^2}$.

**Example 4.3:** Position control system.

We consider the a model for a position control system given as: $G(s) = \frac{1}{s(0.1s+1)}$.

Let a phase-lead controller for the feedback control system be defined as:
$K(s) = K\left(\frac{0.1s+1}{0.02s+1}\right).$

Then, the characteristic polynomial for the closed-loop system is obtained as: $\Delta(s, K) = s(0.02s+1)+K$, or equivalently, $\Delta(s, K) = s^2+50s+50K$.

Let a desired characteristic polynomial be given as: $\Delta_{des}(s) = s^2+50s+1000$.

Then, by comparing the coefficients, we obtain the controller gain as: $K = 20$.

Hence, the dynamic controller for the system is given as: $K(s) = 20(\frac{0.1s+1}{0.02s+1}).$

## 4.2.2 The PID Controller

The PID (proportional-integral-derivate) controller is a general-purpose controller that combines the three basic modes of control, that is, the proportional (P), the derivative (D), and the integral (I) modes.

The PID controller with input $e$ and output $u$ is described by the relation:

$$u(t) = k_p + k_d\frac{d}{dt}e(t) + k_i \int e(t)\,dt$$

or by its Laplace transform equivalent:

$$u(s) = k_p + k_d s + k_i/s.$$

The constants, $\{k_p, k_d, k_i\}$, in the above equations represent the controller gains for the three basic control modes, that is, the proportional, the derivative, and the integral modes (Figure 4.3). Of these, the proportional

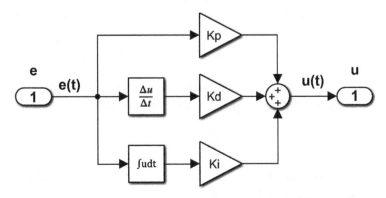

**Figure 4.3**   Three basic control modes represented in the PID controller.

mode mimics the static gain controller that affects the roots of the closed-loop characteristic polynomial. The derivative term, similar to the phase-lead controller, speeds up the system response; and the integral term, similar to the phase-lag controller, improves the steady-state error response of the system.

PID controllers are popular in the industrial settings for process control applications. Often, the controller has to be tuned in response to changes in the process parameters. Industrial process controllers are manually tuned using control knobs on the control boxes, which involves selection of the controller gains: $\{k_p, k_d, k_i\}$. Empirical tuning rules, for example, Ziegler–Nicholas rules (see References), are available and often employed for the tuning of industrial process controllers.

The control system design objectives may require using only a subset of the three basic controller modes. These choices include the proportional control, proportional and derivative (PD) control, and proportional and integral (PI) control, as described below.

**Proportional (P) Controller.** A proportional controller, $K(s) = k_p$, is similar to the static gain controller.

**Proportional–Derivative (PD) Controller.** A PD controller is described by the transfer function:

$$K(s) = k_p + k_d s = k_d \left( s + \frac{k_p}{k_d} \right).$$

A PD controller thus adds a single zero to the loop transfer function that helps speed up the system response. A pure PD controller has the undesirable effect of amplifying high-frequency noise. A filter pole farther (4 or 5 times) to the left of the controller zero is normally added to limit the loop gain at high frequencies to suppress the high-frequency noise.

The modified PD controller is described by the transfer function:

$$K(s) = k_p + \frac{k_d s}{T_f s + 1} = \frac{(k_p T_f + k_d)s + k_p}{T_f s + 1}$$

The modified PD controller adds a zero and a pole to the loop transfer function and is similar to a phase-lead controller, that is:

$$K(s) = \frac{K(s + z_c)}{(s + p_c)}, \quad p_c > z_c.$$

A PD controller, like the phase-lead controller, is employed to improve the transient response of the system.

**Proportional–Integral (PI) Controller.** A PI controller is described by the transfer function:

$$K(s) = k_p + \frac{k_i}{s} = \frac{k_p(s + k_i/k_p)}{s}.$$

The PI controller adds a pole at the origin (i.e., an integrator) and a finite zero to the feedback loop. The controller zero is normally placed close to the origin in the complex $s$-plane. The presence of the integrator in the loop forces the steady-state error to a unit-step input to go to zero; hence, the PI controller is commonly used in designing tracking systems and servomechanisms.

The presence of a pole–zero pair close to the origin adds a closed-loop pole with a large time constant that results in the addition of a relatively slow mode to the system response. This mode normally has a small coefficient and hence does not appreciably degrade the speed of response.

**Proportional–Integral–Derivative (PID) Control.** The PID controller is described by the transfer function:

$$K(s) = k_p + k_d s + \frac{k_i}{s} = \frac{k_d s^2 + k_p s + k_i}{s}.$$

The PID controller includes both PD and PI parts; hence, it adds two zeros and an integrator pole to the loop transfer function. The zero due to the PI part may be located close to the origin; the zero due to the PD part is placed at a suitable location for desired transient response improvement.

A PID controller is alternatively described by a cascade of PI and PD sections, given as:

$$K(s) = \left(k_p + \frac{k_i}{s}\right)(1 + k_d s) \quad \text{or} \quad K(s) = (k_p + k_d s)\left(1 + \frac{k_i}{s}\right).$$

For noise suppression, a first-order lag is added to the derivative term in the PID controller resulting in the modified controller transfer function, given as:

$$K(s) = k_p + \frac{k_i}{s} + \frac{k_d s}{T_f s + 1}$$

The lag has the additional desired effect of making the controller transfer function proper and hence realizable by a combination of a low-pass and high-pass filters.

The PID controller imparts both transient response and steady-state response improvements to the system response. Further, it delivers stability as well as robustness to the closed-loop system.

**PID Controller in MATLAB.** In the MATLAB Control System Toolbox, a PID controller object is created using the "pid" command. The MATLAB software additionally includes a built-in function for the tuning of PID controllers. The relevant commands for creating and tuning PID controllers are given as follows:

```
C0 = pid(Kp,Ki,Kd,Tf);      % create PID controller object
C=pidtune(sys,C0);          % tune PID controller
```

In the above, the argument "sys" denotes a dynamic system object and C0 is a baseline controller. A dynamic system object is created by one of the following MATLAB commands: "tf" (for transfer function), "ss" (for state space), and "frd" (for frequency response design).

The MATLAB PID controller tuning algorithm aims for a $60°$ phase margin, which results in a moderate 8–10% overshoot in the closed-loop system response to a step reference command.

**Example 4.4:** PID controller for the DC motor model.

For a small DC motor, the following parameter values are assumed: $R = 1\,\Omega$, $L = 10\,\text{mH}$, $J = 0.01\,\text{kg} \cdot \text{m}^2$, $b = 0.1\frac{\text{N·s}}{\text{rad}}$, and $k_t = k_b = 0.05$; the transfer function of the DC motor from armature voltage to motor speed is given as:

$$\frac{\omega(s)}{V_a(s)} = \frac{500}{(s + 100)(s + 10) + 25}.$$

We use the MATLAB Control System Toolbox to design and tune a PID controller for the DC motor model and plot the step response of the closed-loop system. The MATLAB commands for this example are given below:

```
G=tf(500,[1 110 1025]);     % define DC motor transfer function
C0=pid(1,1,1,.1);           % create PID controller object
C=pidtune(G,C0)             % tune PID controller for DC motor
T=feedback(C*G,1)           % formulate closed-loop system
step(T),grid                % plot step response
```

The resulting step response of the PID-controlled DC motor is shown in Figure 4.4. We make the following observations from the step response plot:

1. The step response has about 7.5% overshoot.
2. The response settles in about 0.3 s.
3. The steady-state error of the step response is zero.

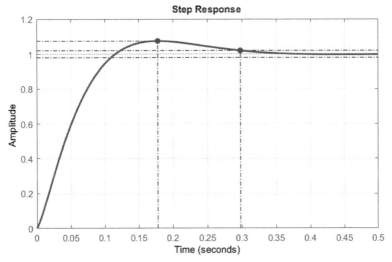

**Figure 4.4** Step response of the DC motor with cascade PID controller.

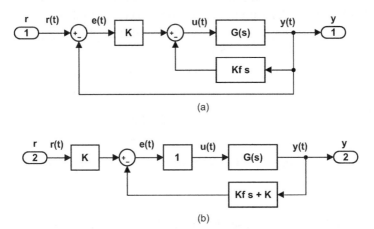

**Figure 4.5** The rate feedback controller: basic rate feedback configuration (a); equivalent PD compensator in the feedback loop (b).

### 4.2.3 Rate Feedback Controllers

In position control applications, rate feedback signal is generally available using a tachometer or a rate gyro and can be employed to further improve the response of the closed-loop system.

The basic rate feedback configuration is shown in Figure 4.5(a), where $k_f$ denotes the rate constant. The closed-loop transfer function for the rate feedback configuration can be obtained by applying Mason's gain rule defined for signal flow graphs.

For simple block diagram configurations involving both output and rate feedback, the gain rule is stated as:

$$\frac{y(s)}{r(s)} = \frac{F_i(s)}{1 - \sum L_i(s)},$$

where $F_i(s)$ denotes the forward path gain from an input node to an output node, and $\sum L_i(s)$ denotes the sum of the loop gains.

**Rate Feedback as PD Controller.** The effective loop gain in the case or rate feedback configuration (Figure 4.5(a)) is computed as:

$$\sum L_i(s) = -G(s)\left(k_f s + K\right) = -KG(s)\left(1 + \frac{k_f}{K}\right).$$

The overall loop gain thus includes the plant transfer function in cascade with a PD controller with the compensator zero located at: $s = -\frac{K}{k_f}$. Hence, rate feedback control configuration is identical to placing a PD controller in the feedback loop (Figure 4.5(b)).

To explore further, let the plant transfer function be given as: $G(s) = \frac{n(s)}{d(s)}$; then, the closed-loop transfer function for the rate feedback configuration is described as:

$$\frac{y(s)}{r(s)} = \frac{Kn(s)}{d(s) + (k_f s + K)n(s)},$$

that is, the resulting closed-loop characteristic polynomial is given as:

$$\Delta(s) = d(s) + (k_f s + K)n(s).$$

**Rate Feedback with Cascade PI as PID Controller.** The rate feedback in conjunction with a cascade PI controller can effectively implement a PID controller. Let the cascade PI controller be defined as: $K(s) = k_p + \frac{k_i}{s}$. Then, the effective loop gain is obtained as (Figure 4.6(a)):

$$\sum L_i(s) = -\left(k_p + \frac{k_i}{s} + k_f s\right) G(s).$$

Hence, the effect of using a cascade PI controller with rate feedback is equivalent to placing a PID controller in the loop (Figure 4.6(b)). Further, let

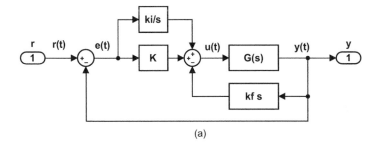

(a)

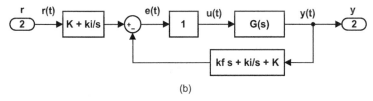

(b)

**Figure 4.6** The rate feedback with cascade PI controller: basic controller configuration (a); equivalent PID compensator in the feedback loop (b).

the plant transfer function be: $G(s) = \frac{n(s)}{d(s)}$; then, the resulting closed-loop characteristic polynomial is given as:

$$\Delta(s) = d(s) + \left(k_p + \frac{k_i}{s} + k_f s\right) n(s).$$

**Example 4.5.** Mass–spring–damper system.

We consider the mass–spring–damper system where the transfer function is given as: $G(s) = \frac{x(s)}{f(s)} = \frac{1}{ms^2 + bs + k}$, and assume the following parameter values: $m = 1, b = 2$, and $k = 2$.

The designs of rate feedback controller and rate feedback with cascade PI controller are given below:

**Rate Feedback Controller Design.** The closed-loop characteristic polynomial for the rate feedback configuration is given as:

$$\Delta(s) = s^2 + (k_f + 2)s + K + 2,$$

where $K$ and $k_f$ are the controller parameters. Let a desired characteristic polynomial be selected as: $\Delta_{des}(s) = (s^2 + 4s + 10)$. Then, by comparing the coefficients, the required controller gains are given as: $K = 8, k_f = 2$.

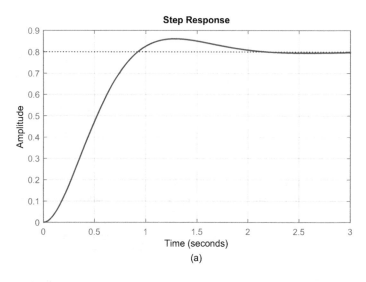

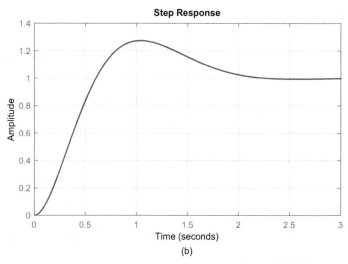

**Figure 4.7** Step responde of the mass–spring–damper system: rate feedback (a); rate feedback with cascade PI (b).

The unit step response of the resulting closed-loop system is shown in Figure 4.7(a).

**Rate Feedback with Cascade PI Controller Design.** The closed-loop characteristic polynomial for the rate feedback configuration with cascade PI controller is given as:

$$\Delta(s) = s^3 + (k_f + 2)s^2 + (k_p + 2)s + k_i$$

The controller gains $\{k_f, k_p, k_i\}$ can be selected for the desired closed-loop performance. Let a desired closed-loop characteristic polynomial is given as: $\Delta_{des}(s) = (s^2 + 4s + 10)(s + 2)$. Then, the controller gains are selected as: $k_f = 4$, $k_p = 20$, and $k_i = 16$. The unit step response of the resulting closed-loop system is shown in Figure 4.7(b).

By comparing the step responses for the two controllers, we make the following observations:

1. The step response of the closed-loop system with only rate feedback controller has about 6.5% overshoot and 20% steady-state tracking error.
2. The step response of the closed-loop system with rate feedback and cascade PI controller has a larger 27% overshoot with zero tracking error. The overshoot can, however, be reduced by moving the real root in the desired characteristic polynomial closer to the origin.
3. In both cases, the unit-step response settles in about 2 s.

The MATLAB script for this example is given below:

```
G=tf(1,[1 2 2]);           % system transfer function
K1=pid(8,0,2);             % PD controller design
T1=8*feedback(G,K1);       % closed-loop system
step(T1),grid              % plot step response
K2=pid(16,20,4);           % PID controller design
K3=pid(16,20,0);           % PI controller design
T2=K3*feedback(G,K2);      % closed-loop system
step(T2),grid              % plot step response
```

## Skill Assessment Questions

Link to the answers:
http://www.riverpublishers.com/book_details.php?book_id=449

1. The model of a hydraulic reservoir level control system is given as: $G(s) = \frac{1}{10s+1}$. The model is connected in a unity-gain feedback configuration with a static controller in the loop.

   (a) Obtain the expression for the closed-loop characteristic polynomial.
   (b) Design a static gain controller for a desired characteristic polynomial: $\Delta_{des}(s) = s + 2$.

2. The model of a position control system is given as: $G(s) = \frac{1}{s(0.2s+1)}$. Assume that the plant is connected in unity-gain feedback configuration with a controller, $K(s)$. Obtain the closed-loop characteristic polynomial for the following controller choices:

   (a) $K(s) = K$
   (b) $K(s) = K\left(\frac{s+5}{s+10}\right)$
   (c) $K(s) = K\left(\frac{s+5}{0.5s+1}\right)$
   (d) $K(s) = K\left(\frac{s+5}{s+10}\right)\left(\frac{s+5}{0.5s+1}\right)$

3. Consider the model of the position control system in Question 2. Assume that the plant is connected in a unity-gain feedback configuration with a static controller in the loop. Design a static gain controller for a desired characteristic polynomial: $\Delta_{des}(s) = s^2 + 5s + 10$.

4. The simplified model of a DC motor with voltage input and position output is described by the following equations: $e_a = R_a i_a + K_b \omega$; $K_t i_a = J\dot{\omega} + D\omega$; $\omega = \dot{\theta}$. The following parameter values are assumed: $R_a = 1$, $J = 0.01$, $D = 0.05$, and $K_t = K_b = 0.1$. The motor is connected in unity-gain feedback configuration with a controller $K(s)$ in the loop. Obtain the closed-loop characteristic polynomial for the following controller choices:

   (a) $K(s) = K$
   (b) $K(s) = K(s + 10)$
   (c) $K(s) = K\left(5 + \frac{1}{s}\right)$
   (d) $K(s) = K\left(5 + s + \frac{1}{s}\right)$

5. Consider the DC motor model in Question 4. Use MATLAB 'pidtune' command to design a PID controller for the motor. Plot the step response of the closed-loop system.

6. A model of human postural dynamic while standing is given as: $G(s) = \frac{0.1}{s^2-9}$. The model is connected in unity-gain feedback configuration with a controller, $K(s)$. Obtain the closed-loop characteristic polynomial for the following controller choices:

   (a) $K(s) = K$
   (b) $K(s) = K\left(\frac{s+a}{s+b}\right)$
   (c) $K(s) = k_p + k_d s + \frac{k_i}{s}$

7. Consider the model of human postural dynamics in Question 6. Use MATLAB 'pidtune' function to design a PID controller for the model. Plot the step response of the closed-loop system.

# 5

---

# Control System Design Objectives

---

## Learning Objectives

1. Characterize the stability of a feedback control system.
2. Characterize performance goals for transient and/or steady-state response improvement.
3. Characterize the disturbance rejection problem in feedback control systems.
4. Determine the sensitivity of the closed-loop system to parameter variations.

The goal of the control system design is to find a controller to condition the input to a physical system in order to obtain a desired behavior at its output, where the physical system may be an automobile, an airplane, a DC motor, an industrial process, etc. In addition, the controlled system is expected to work autonomously without human intervention, much like the cruise control of an automobile.

An automatic control system invariably involves feedback of the output to the input side. The feedback is beneficial in several ways: it permits precise control of the overall system gain and hence its time and frequency responses; it makes the system robust against parameter variations and unmodeled dynamics; and, it can compensate for signal distortions and nonlinearities.

The feedback control system model considered in this chapter (Figure 5.1) includes a plant $G(s)$, an observation model $H(s)$, and a controller, $K(s)$, that suitably modifies the error signal; the latter is obtained by comparing the feedback against a reference signal. The static controller is defined by a constant gain, $K$, that acts as a design parameter and affects the closed-loop behavior of the system.

The static gain controller, however, has its limitations and may be replaced by a dynamic controller, for example, a PID controller. A dynamic

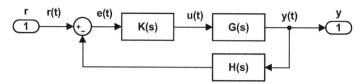

**Figure 5.1** Feedback control system with plant $G(s)$, sensor $H(s)$, and controller represented by $K(s)$.

controller is described by a transfer function, $K(s)$, that contributes to the overall feedback loop transfer function.

The feedback controller design for a given system model is aimed to realize a set of design objectives. The common design objectives include closed-loop system stability, transient response shaping, steady-state error reduction, and sensitivity reduction to disturbance inputs and model parameter variations. These objectives are discussed in the remaining of this chapter.

The control system design objectives involve inherent trade-offs. For example, a static gain controller cannot simultaneously improve the transient response and reduce steady-state error to a unit-step input. Similarly, disturbance rejection and reference tracking pose conflicting design objectives. When faced with such situations, it is not always easy to find the right balance in meeting design objectives. Hence, the control system design is more of an art than an exact science, that is, it relies much on the skills of the designer.

## 5.1 Stability of the Closed-Loop System

Ensuring the closed-loop stability is the first and foremost control system design objective. Even though the physical plant, $G(s)$, may be stable, the presence of feedback can cause the closed-loop system to become unstable. This would happen for high-loop gains when the plant transfer function is at least third order (more on this in Chapter 6).

To explore the closed-loop system stability, we begin by formulating and analyzing the closed-loop characteristic polynomial.

### 5.1.1 Closed-Loop Characteristic Polynomial

The block diagram of the feedback control system with controller in the loop is shown in Figure 5.1. In the diagram, $G(s)$ represents the plant transfer function, $H(s)$ defines the sensor transfer function, and $K(s)$ represents the controller.

The overall system transfer function from input, $r(t)$, to output, $y(t)$, can be obtained by considering the error signal, that is,

$$e = r - Hy = r - KGHe.$$

Further, the output is expressed as:

$$y = KGe = \frac{KGr}{1 + KGH}.$$

The resulting closed-loop transfer function is given as:

$$T(s) = \frac{KG(s)}{1 + KGH(s)}.$$

For sensitivity analysis, we also define an error transfer function, from $r$ to $e$, given as:

$$S(s) = \frac{1}{1 + KGH(s)}.$$

These two transfer functions obey an important identity: $S(s) + T(s) = 1$, which serves as a fundamental constraint in the design of robust control systems.

The determination of stability is based on the closed-loop characteristic polynomial:

$$\Delta(s, K) = 1 + KGH(s).$$

In particular, let $G(s) = \frac{n(s)}{d(s)}$, and let $K(s) = K$ define a static controller; then, the resulting characteristic polynomial is given as:

$$\Delta(s, K) = d(s) + Kn(s).$$

In a more general case, we may replace the static controller with a dynamic one described by: $K(s) = \frac{n_c(s)}{d_c(s)}$. Then, the characteristic polynomial assumes the form:

$$\Delta(s) = d(s)d_c(s) + n(s)n_c(s).$$

In the following, we assume that $\Delta(s)$ is a polynomial of degree $n$, expressed as:

$$\Delta(s) = s^n + a_1 s^{n-1} + \cdots + a_{n-1}s + a_n.$$

### 5.1.2 Stability Determination by Algebraic Methods

The stability determination for the characteristic polynomial uses algebraic methods to characterize its root locations based on the coefficients of the polynomial. A stable (Hurwitz) polynomial has its roots located in the open left-half plane (OLHP).

A necessary condition for stability of the polynomial is that all coefficients of the polynomial are nonzero and are of same sign.

A sufficient condition for polynomial stability is obtained through the application of either of the following equivalent criteria:

**The Hurwitz Criterion.** Consider an $n$th order polynomial, $s^n + a_1 s^{n-1} + \cdots + a_n$. According to the Hurwitz's criterion, the polynomial is stable, that is, it has its roots in the OLHP, if and only if the following determinants have positive values:

$$|a_1|, \quad \begin{vmatrix} a_1 & a_3 \\ 1 & a_2 \end{vmatrix}, \quad \begin{vmatrix} a_1 & a_3 & a_5 \\ 1 & a_2 & a_4 \\ 0 & a_1 & a_3 \end{vmatrix}, \ldots$$

**The Routh's Criterion.** The Routh's criterion involves construction of a Routh's array, where the first two rows in the array are filled by alternating polynomial coefficients.

$$
\begin{array}{c|ccc}
s^n & 1 & a_2 & \cdots \\
s^{n-1} & a_1 & a_3 & \cdots \\
\vdots & b_1 & b_2 & \cdots \\
& c_1 & c_2 & \cdots \\
s^1 & \cdots & & \\
s^0 & \cdots & &
\end{array}
$$

The entries appearing in the third and subsequent rows are computed as follows:

$$b_1 = -\frac{1}{a_1} \begin{vmatrix} 1 & a_3 \\ a_1 & a_2 \end{vmatrix}, \quad b_2 = -\frac{1}{a_3} \begin{vmatrix} a_2 & a_4 \\ a_3 & a_5 \end{vmatrix}, \quad c_1 = -\frac{1}{b_1} \begin{vmatrix} a_1 & a_3 \\ b_1 & b_2 \end{vmatrix}, \text{ etc.}$$

The Routh's stability criterion states that the number of unstable roots of the polynomial equals the number of sign changes in the first column of the array.

**Stability Determination for Low-Order Polynomials.** Low-order polynomials (i.e., polynomials of degree $n = 2, 3$) are often encountered in model-based control system design. The Routh or Hurwitz can be simplified when applied to such polynomials. Using either criterion, the stability conditions for low-order polynomials are given as:

Second-order polynomial ($n = 2$): $a_1 > 0$, $a_2 > 0$.

Third-order polynomial ($n = 3$): $a_1 > 0$, $a_2 > 0$, $a_3 > 0$, $a_1a_2 - a_3 > 0$.

The above stability conditions may be used to determine the range of controller gain, $K$, that would ensure that the roots of the closed-loop characteristic polynomial, $\Delta(s, K)$, lie in the OLHP.

The stability conditions affecting the controller gain are illustrated through the following examples.

**Example 5.1:** A second-order system.

Let $G(s) = \frac{K}{s(s+2)}$, $H(s) = 1$; then, $\Delta(s, K) = s^2 + 2s + K$. By using the above stability criteria, $\Delta(s)$ is stable for $K > 0$.

**Example 5.2:** A third-order system.

Let $\Delta(s, K) = s^3 + 3s^2 + 2s + K$. Then, by using the above stability criteria, $\Delta(s)$ is stable if the following conditions are met: $K > 0$ and $6 - K > 0$. Accordingly, the range of $K$ for closed-loop stablity is given as $0 < K < 6$.

**Example 5.3:** PID controller for simplified model of a DC motor.

The simplified model of a small DC motor is given as: $G(s) = \frac{\theta(s)}{V_a(s)} = \frac{10}{s(s+6)}$.

Let a PID controller be selected as: $K(s) = k_p + k_d s + \frac{k_p}{s}$, and assume that the motor is connected in a unity gain feedback configuration, that is, $H(s) = 1$. Then, the closed-loop characteristic polynomial is given as: $\Delta(s) = d(s)d_c(s) + n(s)n_c(s)$, or,

$$\Delta(s) = s^2(s+6) + 10(k_d s^2 + k_p s + k_i) = s^3 + (6 + 10k_d)s^2 + 10k_p s + 10k_i.$$

The resulting constraints on the PID controller gains to ensure the stability of the third-order polynomial are given as:

$$k_p,\ k_i,\ 6 + 10k_d > 0$$
$$k_p(6 + 10k_d) - k_i > 0.$$

We may choose, for example, $k_p = 1$, $k_i = 1$, $k_d = 1$ to meet the stability requirements.

### 5.1.3 Stability Determination from the Bode Plot

The frequency response of the loop transfer function in a feedback control system can be used to determine the stability of the closed-loop system as follows: given the closed-loop transfer function, $T(s)$, the magnitude of the closed-loop frequency response (in dB) is obtained as:

$$20 \log |T(j\omega)| = 20 \log |KG(j\omega)| - 20 \log |1 + KGH(j\omega)|$$

The closed-loop frequency response will have a discontinuity at the zeros of the closed-loop characteristic polynomial, that is, for $1 + KGH(j\omega) = 0$, or when $KGH(j\omega) = -1$. Hence, the stability of the closed-loop system can be inferred from the frequency response of the loop transfer function: $KGH(j\omega)$, relative to the $1\angle \pm 180°$ point on the complex plane. The stability determination is done using the following relative stability criteria defined on the Bode plot of the frequency response:

**Gain Margin.** It is the factor by which the loop gain, $|KGH(j\omega)|$, can be increased without compromising the closed-loop stability. The gain margin (GM) is computed as:

$$GM = -|KGH(j\omega_{pc})|_{\text{dB}},$$

where $\omega_{pc}$ denotes the phase crossover frequency, defined by $\angle KGH(j\omega) = -180°$.

**Phase Margin.** It is the additional phase angle that can be added to $\angle KGH(j\omega)$ without compromising the closed-loop stability. The phase margin (PM) is computed as:

$$PM = \angle KGH(j\omega_{gc}) + 180°,$$

where $\omega_{gc}$ denotes the gain crossover frequency, defined by $|KGH(j\omega_{gc})|_{\text{dB}} = 0$ dB.

Let the plant transfer function be given as: $G(s) = \frac{n(s)}{d(s)}$, where $n(s)$ and $d(s)$ are, respectively, $m$th and $n$th order polynomials; then, we may observe that the phase crossover becomes relevant when the pole excess, that is, the difference of the number of poles and zeros, exceeds two, that is, $n - m > 2$.

Hence, for low-order plant models ($n \leq 2$), the loop gain, $KGH(s)$, displays an infinite GM as long as the plant transfer function, $G(s)$, is stable.

Similarly, the PM becomes relevant when $|KGH(j0)|_{\text{dB}} > 0$ dB.

Both gain and phase margins are required to be positive to ensure closed-loop stability of the characteristic polynomial. In addition, the phase margin should be adequate to ensure dynamic stability, that is, to ensure that any oscillations in the system response die out in a reasonably fast time.

The relative stability margins can be easily obtained in the MATLAB Control Systems Toolbox by using the "margin" command, as illustrated by the following examples.

**Example 5.4.** DC motor with velocity output.

The model of a small DC motor, with an amplifier with a gain of 3 is given as:

$$\frac{\omega(s)}{V_a(s)} = \frac{1500}{(s+100)(s+10)+25}.$$

The relative stability margins for the model are given as: $GM = \infty; PM = 127°$ (Figure 5.2(a)). These margins were obtained by using the following MATLAB commands:

```
G=tf(1500,[1 110 1025]);      % define plant transfer function
margin(G),grid                % obtain relative stability margins
```

**Example 5.5.** DC motor with position output.

The model of a small DC motor including an amplifier with a gain of 3 is used in a position control application, where the plant transfer function is given as:

$$\frac{\theta(s)}{V_a(s)} = \frac{1500}{s[(s+100)(s+10)+25]}.$$

The relative stability margins, obtained by using the following MATLAB commands, are given as: $GM = 37.5$ dB; $PM = 81.1°$ (Figure 5.2(b)).

```
G=tf(1500,[1 110 1025 0]);    % define plant transfer function
margin(G),grid                % obtain relative stability margins
```

## 5.2 Transient Response Improvement

The closed-loop system should be stable, that is, its transient response should die out with time; in addition, the closed-loop poles must have adequate damping to ensure dynamic stability, so that any oscillations in the output response die out in a reasonably short time.

The transient response of a dynamic system constitutes a weighted sum of system's natural response modes, which are defined by the roots of the

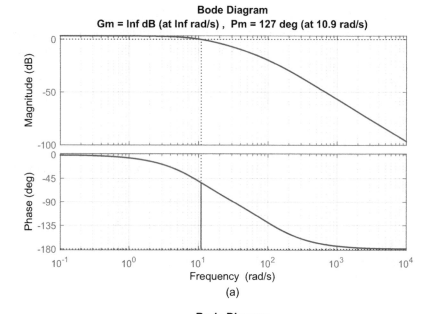

(a)

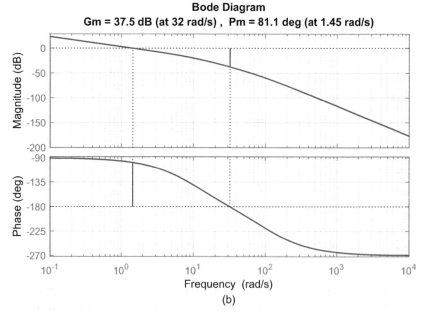

(b)

**Figure 5.2**    Determination of relative stability margins on the Bode plot of the loop transfer function; (a) $KGH(s) = \frac{1500}{s^2+110s+1025}$; (b) $KGH(s) = \frac{1500}{s(s^2+110s+1025)}$.

closed-loop characteristic polynomial (Section 2.1.2). These roots can be real or complex, distinct or repeated. Accordingly, the system natural response modes are characterized as follows:

**Real and Distinct Roots.** Assume that the $n$ roots of the characteristic polynomial are real and distinct, so that the polynomial can be factored as: $\Delta(s) = (s - p_1)(s - p_2) \dots (s - p_n)$. Then, the system natural response modes are given as: $\{e^{p_1 t}, e^{p_2 t}, \dots, e^{p_n t}\}$.

**Real and Repeated Roots.** Assume that the characteristic polynomial has real and repeated roots, that is, it contains factors of the form: $\Delta_m(s) = (s - s_1)^m$. Then, their contribution to system natural response modes is given as: $\{e^{s_1 t}, t e^{s_1 t}, \dots, t^{m-1} e^{s_1 t}\}$.

**Complex Roots.** In this case, each quadratic factor with complex roots can be written as: $\Delta_c(s) = (s + \sigma)^2 + \omega^2$. The corresponding natural response modes are given as: $\{e^{-\sigma t} \cos \omega t, e^{-\sigma t} \sin \omega t\}$.

**General Expression.** To generalize, we assume that for an $n$th order system, the system natural response modes are given as: $\phi_k(t), k = 1, \dots, n$. Then, using arbitrary constants, $c_k$, the transient response of the closed-loop system is given as: $y(t) = \sum_{k=1}^{n} c_k \phi_k(t)$.

## 5.2.1 System Design Specifications

The system design specifications are commonly stated in terms of one or more of the following parameters with reference to the step response of the closed-loop system (Figure 5.3): (a) rise time ($t_r$), (b) settling time ($t_s$), (c) damping ratio ($\zeta$), and (d) percentage overshoot ($\%OS$).

The rise time is defined in multiple ways: when system response asymptotically reaches its steady-state value, the rise time is defined as the time to reach from 10% to 90% of the final value. For a system that has an overshoot in the step response, the rise time is defined as the time when the step response first reaches 100% of its final value.

The rise time is related to the closed-loop system bandwidth, $\omega_B$, through $\omega_B t_r \approx 1$.

The settling time is defined as the time when the system step response settles to within 1% (2%, according to some authors) of its final value. Since $1 - e^{-4.5} \cong 0.99$, we may approximate the settling time as: $t_s \cong 4.5\tau$, where $\tau$ is the time constant for the dominant real or complex roots, that is, the roots closest to the imaginary axis.

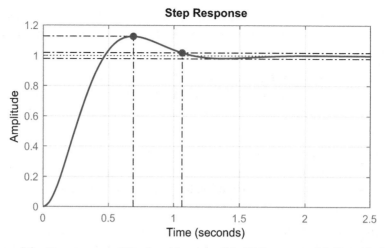

**Figure 5.3**  Step response of the closed-loop simplified DC motor model (Example 5.6).

**Prototype Second-Order System.** In order to relate the design specifications to the system step response, we consider the closed-loop system represented by a prototype second-order transfer function, given as:

$$T(s) = \frac{\omega_n^2}{s^2 + 2\zeta\omega_n s + \omega_n^2}.$$

The denominator polynomial above is assumed to have complex roots at: $s = -\zeta\omega_n \pm j\omega_d$, where $\omega_d = \omega_n\sqrt{1 - \zeta^2}$ represents the damped natural frequency, that is, the frequency of oscillations of the system output.

For the prototype second-order system, the transient response quality indicators are defined as (Section 2.2):

$$t_r = \frac{\pi - \phi}{\omega_d}, \quad \phi = \tan^{-1}\frac{\sqrt{1 - \zeta^2}}{\zeta}$$

$$t_s \cong \frac{4}{|\sigma|}, \quad \sigma = Re[s]$$

$$\%OS = 100\,e^{-\pi\zeta\omega_n/\omega_d}, \quad \zeta = \zeta(\%OS).$$

In particular, the step response overshoot and the normalized rise time, $\omega_n t_r$, are functions of the damping ratio, $\zeta$. A short table of $\zeta$ versus the $\%OS$ and $\omega_n t_r$ is given below:

| $\zeta$ | 0.9 | 0.8 | 0.7 | 0.6 | 0.5 |
|---|---|---|---|---|---|
| $\%OS$ | 0.2% | 1.5% | 4.6% | 9.5% | 16.3% |
| $\omega_n t_r$ | 5.54 | 4.18 | 3.3 | 2.76 | 2.42 |

A low ($\leq 10\%$) overshoot in the step response is often desired that translates into $\zeta \geq 0.6$.

**Example 5.6:** Simplified model of a DC motor.

The simplified model of a small DC motor is given as: $G(s) = \frac{\theta(s)}{V_a(s)} = \frac{10}{s(s+6)}$. Using a static gain controller, the closed-loop transfer function is obtained as: $T(s) = \frac{10K}{s^2+6s+10K}$.

Then, for $K = 3$, the closed loop roots are located at: $s = -3 \pm j4.58$. The resulting damping ratio is: $\zeta = 0.55$, which results in a step overshoot of 13%.

The system rise time is $t_r \cong 0.47s$ with $\omega_n t_r \cong 3$, and the settling time is $t_s \cong 1.06\ s$ (Figure 5.3).

The following commands from the MATLAB Control Systems Toolbox were used to obtain the step response. The peak time and settling time characteristics were added by right clicking on the plot.

```
G=tf(30,[1 6 30]);      % define system transfer function
step(G),grid            % plot step response
```

## 5.2.2 The Desired Characteristic Polynomial

The controller design specifications given in terms of rise time ($t_r$), settling time ($t_s$), damping ratio ($\zeta$), and/or percentage overshoot ($\%OS$) of system's step response can be used as a guide to choose the root locations for a desired closed-loop characteristic polynomial.

Using the prototype second-order system, we can translate the design specifications, specified as a subset of $\{t_r, t_s, \zeta, \%OS\}$, into the desired real part ($\sigma$), magnitude ($\omega_n$), and angle ($\theta = \cos^{-1}\zeta$) of the closed-loop root locations. Accordingly, we may choose:

$$\sigma \geq \frac{4.5}{t_s}, \quad \omega_n \geq \frac{2}{t_r}, \quad \theta = \cos^{-1}\zeta$$

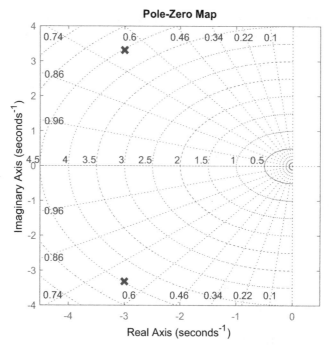

**Figure 5.4**   Pole locations for the desired characteristic polynomial (Example 5.7).

**Example 5.7:** Assume that the design specifications in terms of desired step response are given as: $t_r \leq 0.5s$, $t_s \leq 3.0s$, and $\%OS \leq 10\%$. These specifications translate into the following constraints on the closed-loop pole locations: $\sigma \geq 1.5$, $\omega_n \geq 4$, and $\zeta \geq 0.6$.

Then, we may choose, for example, the desired closed-loop characteristic polynomial as: $\Delta_{des}(s) = (s + \sigma)^2 + \omega_d^2 = s^2 + 6s + 20$.

In the MATLAB Control Systems Toolbox, the constant $\omega_n$ circles and constant $\zeta$ lines can be displayed in the form of a grid. For the above example, this is done using the following MATLAB commands (Figure 5.4):

```
pzmap(tf(1,[1 6 20])), grid    % show grid for control system
axis([-4.5 .5 -4 4])           % adjust axes ranges
```

**Controller Gain Selection.** Since the closed-loop characteristic polynomial includes the static controller gain as a parameter, we may choose the controller gain to satisfy the performance objectives by comparing the coefficients with those of the desired characteristic polynomial. This is illustrated by the following examples.

**Example 5.8:** Static gain controller.

Let $\Delta(s, K) = s^2 + 2s + K$ represent the characteristic polynomial for a control system with static controller, where the objective is to choose $K$ for $\zeta = 0.7$. For the choice of $\zeta$, a desired characteristic polynomial may be given as: $\Delta_{des}(s) = s^2 + 2s + 2$. Hence, we may choose $K = 2$ for the desired damping ration.

We may note that this choice would result in: $t_r \cong 1.5s$ and $t_s \cong 4.5s$.

**Example 5.9:** Rate feedback controller.

Let $\Delta(s, K, K_1) = s^2 + K_1 s + K$ represent the characteristic polynomial for the rate feedback controller configuration. The objective is to choose $K, K_1$ for $\%OS \leq 5\%$ ($\zeta = 0.7$) and $t_s \leq 2.5s$. A desired characteristic polynomial for $\zeta = 0.7$ and $\zeta \omega_n = 2$ is given as: $\Delta_{des}(s) = (s + 2)^2 + 2^2$. Accordingly, $K_1 = 4, K = 8$.

### 5.2.3 Optimal Performance Indices

An alternate way to characterize the transient response of the closed-loop system, and to choose a desired characteristic polynomial, is to define a performance index and choose the controller gain $K$ that minimizes that index. Any such performance index includes the whole of the transient response component of the system response.

Toward this end, three popular performance indices have been defined as:

$$\text{IAE:} \int_0^{t_s} |e(t)|\, dt$$

$$\text{ISE:} \int_0^{t_s} |e(t)|^2\, dt$$

$$\text{ITAE:} \int_0^{t_s} t|e(t)|\, dt.$$

The ITAE index, in particular, has been commonly used to evaluate the system performance in industrial process control. Using a second-order system with a step input ($r(s) = \frac{1}{s}$), minimization of the ITAE index results in an optimal design with $\zeta = 0.7$ ($\%OS = 4.6\%$).

The optimum coefficients of the desired characteristic polynomial for low-order systems based on the ITAE index are given below:

$$s + \omega_n$$

$$s^2 + 1.4\,\omega_n s + \omega_n^2$$

$$s^3 + 1.75\,\omega_n s^2 + 2.15\,\omega_n^2 s + \omega_n^3$$

$$s^4 + 2.1\,\omega_n s^3 + 3.4\,\omega_n^2 s^2 + 2.7\,\omega_n^3 s + \omega_n^3.$$

These coefficients are used as a guide to select the desired characteristic polynomial in controller design. This is illustrated by the following example.

**Example 5.10:** Industrial process control.

The first-order plus dead time (FOPDT) model of an industrial process is given as: $G(s) = \frac{e^{-s}}{s+1}$. We desire to select a controller gain to control the process in a unity feedback gain configuration in accordance with ITAE index.

In order to obtain a rational transfer function, we employ first-order Pade approximation of the delay term: $e^{-s} \cong \frac{2-s}{2+s}$, to write $G(s) \cong \frac{2-s}{(s+1)(s+2)}$.

The resulting characteristic polynomial for a unity feedback gain control system is given as: $\Delta(s) = (s+1)(s+2) + K(2-s)$.

Using ITAE index, let the desired second-order polynomial be: $\Delta(s) = s^2 + 1.4\,\omega_n s + \omega_n^2$. Then, by comparing the coefficients, we obtain a pair of equations that are solved for a positive value of $\omega_n = 1.76\frac{\text{rad}}{\text{s}}$. The desired controller gain is obtained as: $K = 0.54$.

## 5.3 Steady-State Error Improvement

The steady-state response of the closed-loop system is required to have a low-tracking error in response to a constant (i.e., unit step) or linearly varying (i.e., unit ramp) reference input. The desired tracking error tolerance may be specified as percentage of the reference input magnitude. We may additionally require that the system response to a constant disturbance input stays small.

### 5.3.1 The Steady-State Error

Let $T(s)$ denote the closed-loop transfer function; then $y(s) = T(s)r(s)$, and the resulting tracking error is specified as:

$$e(s) = (1 - T(s)r(s)).$$

To characterize the tracking error, we may consider a unity gain feedback system (i.e., let $H(s) = 1$). Then, $e(s) = r(s) - y(s) = r(s) - KGe(s)$.

The tracking error, $e(s)$, in response to a reference signal, $r(s)$, is given as:

$$e(s) = \frac{1}{1 + KG(s)} r(s).$$

Then, using the final-value theorem (FVT), the steady-state tracking error is computed as:

$$e(\infty) = \lim_{s \to 0} se(s).$$

In particular, for a step reference signal, $(r(s) = \frac{1}{s})$, it is given as:

$$e(\infty) = \frac{1}{1 + KG(0)}.$$

Whereas, for a ramp reference signal, $(r(s) = \frac{1}{s^2})$, it is given as:

$$e(\infty) = \frac{1}{sKG(s)|_{s=0}}.$$

## 5.3.2 System Error Constants

Traditionally, the steady-state tracking error for a unity-gain feedback control system is characterized in terms of system position and velocity error constants, which are defined with reference to the dc gain of the feedback loop as:

$$K_p = \lim_{s \to 0} KG(s),$$
$$K_v = \lim_{s \to 0} sKG(s).$$

In terms of the above error constants, the steady-state tracking error to a step $(r(s) = \frac{1}{s})$ or a ramp $(r(s) = \frac{1}{s^2})$ input is evaluated as:

$$e(\infty)|_{step} = \frac{1}{1 + K_p}$$

$$e(\infty)|_{ramp} = \frac{1}{K_v}.$$

Integral control, that is, $K(s) = \frac{k_i}{s}$, is often employed to reduce the steady-state error. The presence of an integrator in the loop forces $K_p \to \infty$, and hence, the steady-state error to a step input goes to zero, as shown in the following example.

**Example 5.11:** A second-order system.

Let $KG(s) = \frac{K}{s(s+2)}$. Then, $K_p = \infty$, $K_v = \frac{K}{2}$.

Therefore, $e(\infty)|_{step} = 0$, $e(\infty)|_{ramp} = \frac{2}{K}$.

**Example 5.12:** DC motor model.

We consider the model of a small DC motor, where the following component values are assumed: $R = 1\ \Omega$, $L = 10$ mH, $J = 0.01$ kgm$^2$, $b = 0.1$ $\frac{\text{Ns}}{\text{rad}}$, $k_t = k_b = 0.05$.

The transfer function for the DC motor is given as: $G(s) = \frac{500}{s^2 + 110s + 1025}$, so that $G(0) \cong 0.5$, $K_p = 0.5K$, and $K_v = 0$. Hence, $e(\infty)|_{step} = \frac{2}{K}$, $e(\infty)|_{ramp} = \infty$.

### 5.3.3 Steady-State Error to Ramp Input

Assuming that the feedback loop contains an integrator, so the steady-state error to a step input is zero, an alternate expression for the steady-state error to a ramp input can be developed. Let the steady-state error to a ramp be expressed in terms of the closed-loop transfer function as:

$$e(s) = [1 - T(s)]r(s); \quad r(s) = \frac{1}{s^2}.$$

The steady-state error is found by the application of FVT as:

$$e(\infty)|_{ramp} = \lim_{s \to 0} \frac{1 - T(s)}{s}.$$

Further, by using the L' Hospital's rule, we have:

$$e(\infty)|_{ramp} = \lim_{s \to 0} \left( -\frac{dT(s)}{ds} \right).$$

To evaluate the RHS, we use the natural logarithm to write:

$$\frac{d}{ds} \ln T(s) = \frac{1}{T(s)} \frac{dT(s)}{ds}.$$

Since $\lim_{s \to 0} T(s) = 1$ by the integrator assumption, we have:

$$\lim_{s \to 0} \frac{d}{ds} \ln T(s) = \lim_{s \to 0} \frac{dT(s)}{ds}.$$

To proceed further, we assume that $T(s)$ is expressed in the factored form as:

$$T(s) = \frac{K(s - z_1) \ldots (s - z_m)}{(s - p_1) \ldots (s - p_n)}.$$

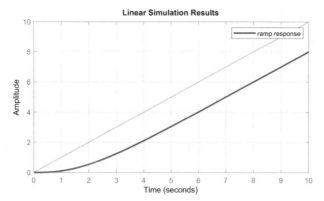

**Figure 5.5** Steady-state error to a ramp input (Example 5.13).

Then, we have:

$$\ln T(s) = \ln K + \sum_{ln}(s - z_i) - \sum_{ln}(s - p_i)$$

$$\frac{d}{ds}\ln T(s) = \ln K + \sum \frac{1}{s - z_i} - \sum \frac{1}{s - p_i}.$$

It follows that:

$$e(\infty)|_{ramp} = \sum \frac{1}{p_i} - \sum \frac{1}{z_i},$$

where $z_i$ are the closed-loop zeros (same as open loop zeros), and $p_i$ are the closed-loop poles.

**Example 5.13:** Steady-state error to ramp input.

Let $KG(s) = \frac{K}{s(s+2)}$; then, the closed-loop transfer function is: $T(s) = \frac{K}{s(s+2)+K}$. Assuming that the controller gain is selected as: $K = 1$, the closed-loop poles are located at: $s_{1,2} = -1, -1$. The steady-state error to a ramp input is given as: $e(\infty)|_{ramp} = \sum \frac{1}{p_i} = -2$.

The above result can be verified by computer simulation of the closed-loop system response using the following commands from MATLAB Control Systems Toolbox (Figure 5.5):

```
T=tf(1,[1 2 1]);              % define system transfer function
lsim(T,0:.01:10,0:.01:10),grid % simulate the ramp response legend
                               ('ramp response')
```

## 5.4 Disturbance Rejection

Disturbance inputs are unavoidable in the operation of physical systems. Common examples of disturbance inputs include road bumps while driving, turbulence in airplanes, machinery vibrations in mechanical plants, etc.

In order to characterize the effects of adding a disturbance input to the feedback control system, we consider a modified block diagram (Figure 5.6).

Let $r(t)$ denote a reference input, and $d(t)$ denote a disturbance input; then, the system output in the presence of both the reference and disturbance inputs is given as:

$$y(s) = \frac{KG(s)}{1 + KGH(s)}r(s) + \frac{G(s)}{1 + KGH(s)}d(s)$$

In particular, we focus on the unity feedback gain configuration $(H(s) = 1)$; then, the tracking error, $e(s)$, is computed as:

$$e(s) = \frac{1}{1 + KG(s)}r(s) - \frac{G(s)}{1 + KG(s)}d(s).$$

Using the FVT, the steady-state error to a step reference $(r(s) = 1/s)$ in the presence of a disturbance input $(d(s) = 1/s)$ is computed as:

$$e(\infty) = \frac{1}{1 + K_p}r(\infty) - \frac{G(0)}{1 + K_p}d(\infty).$$

In order to reduce the effects of a constant input disturbance on the output, the loop gain should be large. A large loop gain also reduces steady-state error to a step reference input. The control system designer can affect the loop gain by choosing a large $K$ in the case of a static controller.

In a more general case of tracking a time-varying reference input, the loop gain should be large over the frequency band represented in the

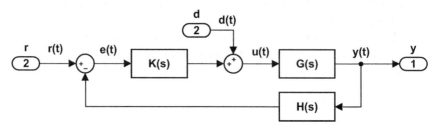

**Figure 5.6**   A feedback control system with reference and disturbance inputs.

reference signal (low frequency restriction). Loop gain should also be high in the frequency band of the disturbance input (high frequency restriction). In between, the loop gain should have a moderate slope around the gain crossover frequency.

A large controller gain, however, generates a large magnitude input signal to the plant, which may cause saturation in the actuator devices (amplifiers, mechanical actuators, etc.). Thus, the control effort needs to be limited. This objective is explicitly formulated in linear optimal control problems (see References).

A large loop gain also tends to reduce damping and undermine the closed-loop stability. With the static gain controller, a design trade-off exists between having adequate stability margins and achieving good disturbance rejection. More complex controller models may be employed to ensure that the competing design objectives are suitably met (Chapter 6).

**Internal Model Control.** To analyze the control requirements for simultaneous tracking and disturbance rejection, let $G(s) = \frac{n_p(s)}{d_p(s)}$ represent the plant, and let $K(s) = \frac{n_c(s)}{d_c(s)}$ represent the controller. Then, the plant output in the presence of referece and disturbance inputs is given as:

$$y(s) = \frac{d_p(s)d_c(s)}{n_p(s)n_c(s) + d_p(s)d_c(s)}r(s) - \frac{n_p(s)d_c(s)}{n_p(s)n_c(s) + d_p(s)d_c(s)}d(s).$$

Then, the requirements for asymptotic tracking and disturbance rejection are given as follows:

**Asymptotic tracking.** For asymptotic tracking, $d_p(s)d_c(s)$ should contain the unstable poles of $r(s)$. As an example, an integrator in the feedback loop ensures zero steady-state error to a constant reference input.

**Disturbance Rejection.** For disturbance rejection, $n_p(s)d_c(s)$ should contain the unstable poles of $d(s)$. For example, a notch filter centered at 60 Hz helps remove the powerline noise from the signal.

**Example 5.14:** A small DC motor.

Consider the model of a small DC motor, where the following component values are assumed: $R = 1\ \Omega$, $L = 10$ mH, $J = 0.01$ kgm$^2$, $b = 0.1\ \frac{\text{Ns}}{\text{rad}}$, $k_t = k_b = 0.05$.

The transfer function of the DC motor is: $G(s) = \frac{500}{s^2+110s+1025}$, so that the DC gain of the system is: $G(0) \cong 0.5$, and the position error constant is:

$K_p = 0.5K$. Then, for a step reference input and/or step disturbance,

$$e(\infty) = \frac{1}{1 + 0.5K} r(\infty) - \frac{0.5}{1 + 0.5K} d(\infty).$$

We may choose a large $K$ to reduce the steady-state tracking error to a reference input, as well as improve the disturbance rejection. A large value of $K$, however, reduces damping in the closed-loop system and results in an oscillatory response of the DC motor. The closed-loop characteristic polynomial is given as: $\Delta(s, K) = s^2 + 110s + 1025 + 500K$.

The resulting damping ratio is computed as: $\zeta = \frac{55}{\sqrt{1025 + 500K}}$.

In order to limit the damping to, say $\zeta \leq 0.6$, the controller gain is limited to: $K \leq 14.75$. We may choose, for example, $K = 14$, which results in: $e(\infty) = \frac{1}{8} r(\infty) - \frac{1}{16} d(\infty)$.

## 5.5 Sensitivity and Robustness

The robustness in a control system refers to its ability to withstand parameter variations, and/or unmodeled dynamics in the plant transfer function, and still maintain the stability and performance goals. Robustness to parameter variations is characterized in terms of the sensitivity of the closed-loop transfer function $T(s)$ to variation in one or more of the plant parameters.

**System Sensitivity Function.** The system sensitivity function is defined as the ratio of the percentage change in the closed-loop transfer function to a percentage change in the plant transfer function, that is:

$$S_G^T = \frac{\partial T/T}{\partial G/G} = \frac{\partial T}{\partial G} \frac{G}{T} = \frac{1}{1 + KGH(s)}.$$

We note that the system sensitivity function matches the error transfer function, that is:

$$\frac{e(s)}{r(s)} = \frac{1}{1 + KGH(s)} = S(s).$$

Hence, increasing loop gain in a feedback control system by choosing a larger controller gain $K$ simultaneously helps reduce its sensitivity to parameter variations. Further, the sensitivity function is frequency-dependent. Hence, low sensitivity in the desired frequency band can be aimed.

The sensitivity of the closed-loop transfer function to percentage change in some parameter $\alpha$ that is represented in the plant transfer function $G(s)$ is given as:

$$S_\alpha^T = S_G^T S_\alpha^G.$$

The sensitivity and robustness calculations are illustrated via the following examples.

**Example 5.15:** parameter sensitivity.

Let $KG(s) = \frac{K}{s(s+a)}$, $H(s) = 1$, where the parameter $a$ may vary over a range of values. Then,

$$T(s) = \frac{K}{s(s+a)+K}; \quad S_G^T = \frac{s(s+a)}{s(s+a)+K};$$
$$S_a^G = -\frac{a}{s+a}; \quad S_a^T = -\frac{as}{s(s+a)+K}.$$

Thus, increasing the loop gain (i.e., by choosing a large $K$) reduces the sensitivity of $T(s)$ to variations in $a$.

**Example 5.16:** A small DC motor.

Consider the model of a small DC motor, where the following component values are assumed: $R = 1$ $\Omega$, $L = 10$ mH, $J = 0.01$ kg m$^2$, $b = 0.1$ $\frac{\text{Ns}}{\text{rad}}$, $k_t = k_b = 0.05$.

The transfer function of the DC motor is given as: $G(s) = \frac{k_t}{(Js+b)(Ls+R)+k_t k_b} = \frac{500}{s^2+110s+1025}$.

Then, $S_G^T = \frac{1}{1+0.5K}$.

Next, we illustrate how we can use the MATLAB Symbolic Math Toolbox to find the sensitivity of the closed-loop transfer function, $T(s)$, to torque constant, $k_t$, where $S_{k_t}^T = S_G^T S_{k_t}^G$.

The MATLAB script is given as follows:

```
R=1; L=.01; J=.01; b=.1; kb=.05;    % define model parameters
syms s K kt                          % define symbolic variables
G(s)=kt/((J*s+b)*(L*s+R)+kt*kb);     % DC motor transfer function
dG=diff(G,kt);                       % differentiate transfer function
S(s)=1/(1+K*G);                      % DC motor sensitivity function
Skt=S*dG*kt/G;                       % CL sensitivity function
kt=.05;                              % define nominal value
subs(Skt)                               % substitute in the expression
simplify(ans)                        % simplify the result
```

The answer is: $S_{k_t}^T = (s^2 + 110s + 1000)/(s^2 + 110s + 500K + 1025)$.

## Skill Assessment Questions

Link to the answers:
http://www.riverpublishers.com/book_details.php?book_id=449

1. The characteristic polynomial of a closed-loop system is given as: $s(s+1)(s+2) + K = 0$. Find the range of $K$ for stability.

2. Use the proto-type second-order system to determine the desired characteristic polynomial for the following design objectives:

    (a) $\omega_n = 1\frac{\text{rad}}{\text{s}}; \zeta = 0.8$

    (b) $\omega_n = 1\frac{\text{rad}}{\text{s}}; \%OS = 5\%$

    (c) $t_r = 0.5s; \zeta = 0.7$

    (d) $t_s = 2s; \%OS = 5\%$

3. Find the steady-state error to a unit-step input for the following transfer functions. For (b) and (d), also find the steady-state error to a ramp input.

    (a) $G(s) = \frac{1}{s+1}$

    (b) $G(s) = \frac{1}{s(s+1)}$

    (c) $G(s) = \frac{1}{(s+1)(s+2)}$

    (d) $G(s) = \frac{s+1}{s(s+2)}$

4. Assume that a plant with $KG(s) = \frac{25K}{s(s+2)(s+25)}$ is connected in unity gain feedback configuration.

    (a) Express the closed-loop characteristic equation and find the range of $K$ for stability.

    (b) Find the steady-state error to: (i) step input, (ii) step disturbance, (iii) ramp input.

5. Let $KG(s) = \frac{K(s+a)}{s(s+1)}$, $H(s) = 1$; find the sensitivity $S_a^T$ of the closed-loop system to variations in $a$.

6. Consider the model of human postural dynamics described as an inverted pendulum: $G(s) = \frac{1}{(I+mh^2)}\frac{1}{(s^2-\Omega^2)}$, where $m$ is the mass, $h$ is the height of center of mass, $I$ is the moment of inertia about center of mass, $g$ is the gravitational constant, and $\Omega^2 = \frac{mgh}{I+mh^2}$. Assume that the control system is connected in a unity-gain feedback configuration. Find the sensitivity of the closed-loop transfer function to variations in $m, I,$ and $h$.

# 6

# Control System Design with Root Locus

## Learning Objectives

1. Sketch the root locus of a given loop transfer function with respect to the controller gain $K$.
2. Use root locus to design static output feedback controller for a given plant transfer function.
3. Use root locus to design first-order dynamic controllers for transient and/or steady-state response improvement.
4. Realize the controller design using analog operational amplifier circuits.

Beginning with this chapter, we discuss the techniques for designing controllers for feedback control systems. The various controller design techniques include the root locus (RL) technique (covered here), full state feedback design (covered in Chapters 7 and 8), and frequency response design (covered in Chapter 9). This chapter focuses on the design of control systems using the RL technique.

The structure of the feedback control system for the RL design (Figure 6.1) includes the process to be controlled (the plant), a sensor that measures the output, and a cascade controller. The plant and the sensor are described in terms of transfer function models, $G(s)$ and $H(s)$, respectively. The controller, represented by static gain $K(s) = K$, multiplies the error signal. The resulting feedback loop transfer function is given as: $KGH(s)$.

We focus on the selection of static gain, $K$, as a parameter in the closed-loop characteristic polynomial, where $K$ takes on positive values. The root locus technique allows graphical representation of achievable closed-loop pole locations. The controller gain is selected based on its ability to meet the design specifications.

133

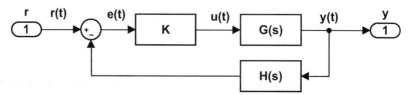

**Figure 6.1** Feedback control system with plant $G(s)$, sensor $H(s)$, and a static gain controller represented by $K$.

The control system is designed for closed-loop stability and desired transient and/or steady-state response improvements. The desired transient response improvements are measured in terms of step response parameters ($t_r$, $t_s$, $\%OS$, etc.), while the steady-state tracking error requirements are given in terms of the relevant error constants ($K_p$, $K_v$).

The root locus plot of a given loop transfer function, $KGH(s)$, constitutes $n$ branches in the complex $s$-plane, where $n$ is the order of the denominator polynomial. These branches commence at the open-loop (OL) poles and proceed towards finite zeros of the loop transfer function, or asymptotically toward infinity. All root locus plots bear common characteristics that are referred as root locus rules. These describe the location of real-axis locus, break points, asymptote directions, etc.

The root locus plots for simple loop transfer functions can often be sketched by hand. More commonly, they are drawn using computer software like MATLAB. Having obtained the root locus plot, the controller gain for a desired root location can be computed by evaluating the loop transfer function at that location. In MATLAB, the controller gain can be read by clicking on the RL plot.

The root locus design technique can be extended to dynamic controllers of phase-lead, phase-lag, lead–lag, PD, PI, and PID types. The dynamic controller introduces poles and zeros that modify the original RL to bring about the desired improvements in design. With the availability of a rate sensors (rate gyroscope and tachometer), rate feedback design that includes the design of inner and outer feedback loops may be considered.

The static and dynamic controllers designed for process improvement can be implemented with electronic circuits built with operational amplifiers, resistors, and capacitors. The realization of the controller transfer function in terms of electronic circuits is discussed toward the end of the chapter.

## 6.1 The Root Locus

The root locus (RL) constitutes a graph of the closed-loop root locations, that is, the roots of the closed-loop characteristic polynomial, with variation in the controller gain. In order to develop the RL plot, we consider a feedback control system (Figure 6.1) with a static gain controller $K$, where $G(s)$ is the plant transfer function and $H(s)$ is the sensor transfer function. We will assume $H(s) = 1$, unless otherwise specified.

### 6.1.1 Roots of the Characteristic Polynomial

The loop transfer function represents the gain along the feedback loop from the output of the summer along the feedback path back to the summer. The loop gain includes the controller, the plant, and the sensor and is given as: $L(s) = KG(s)H(s) = KGH(s)$.

The characteristic polynomial of the closed-loop system is defined in terms of loop transfer function as:

$$\Delta(s, K) = 1 + KGH(s).$$

The roots of $\Delta(s, K)$ vary with the controller gain, $K$, and can be plotted in the complex plane for the various values of $K$. The loci of all such roots locations, as $K$ varies from $0 \to \infty$, constitute the root locus for the characteristic polynomial in the feedback control system.

A generalized RL is an extension of regular RL that constitutes the loci of all root locations of the characteristic polynomial for $K \in (-\infty, \infty)$ and is obtained by extending the RL to the negative values of $K$.

The variation of the closed-loop roots by varying controller gain, $K$, is illustrated in the following examples.

**Example 6.1:** Let $G(s) = \frac{1}{s+1}$, $H(s) = 1$. Then, the loop gain is: $KG(s) = \frac{K}{s+1}$. The closed-loop characteristic polynomial is given as: $\Delta(s) = s+1+K$. The polynomial has a single root at $s_1 = -1 - K$. Hence, the root locus, as $K$ varies from 0 to $\infty$, traces a line that proceeds from $\sigma = -1$ along the negative real-axis to infinity (Figure 6.2(a)).

**Example 6.2:** Let $G(s) = \frac{1}{s(s+2)}$, $H(s) = 1$. Then, the loop gain is: $KG(s) = \frac{K}{s(s+2)}$. The closed-loop characteristic polynomial is given as: $\Delta(s, K) = s^2 + 2s + K$. The roots of the characteristic polynomial are real for $K \leq 1$ and are complex for $K > 1$, where the complex roots are given as:

$s_{1,2} = -1 \pm \sqrt{1-K}$. The closed-loop roots for a range of $K$ are tabulated below.

| $K$ | Roots of $\Delta(s, K)$ |
|-----|-------------------------|
| 0   | $0, -2$ |
| 0.5 | $-0.29, \ -1.71$ |
| 1.0 | $-1, -1$ |
| 1.5 | $-1 \pm j0.71$ |
| 2.0 | $-1 \pm j1$ |
| 2.5 | $-1 \pm j1.22$ |
| 3.0 | $-1 \pm j1.41$ |

The loci of these roots, as $K$ varies from $0 \to \infty$, comprise two branches that commence at the open-loop (OL) poles located at $\{0, -2\}$, proceed inward along the real-axis, meet in the middle at $\sigma = -1$, then split and extend along the $\sigma = -1$ line in both upward and downward directions to $s = -1 \pm j\infty$. These directions are called the RL asymptotes. The RL plot is shown in Figure 6.2(b).

### 6.1.2 Root Locus Rules

All RL plots share common properties, which are referred as the root locus rules. The prominent rules are described below:

1. The RL has $n$ branches, where $n$ equals the order of the denominator polynomial in the loop transfer function $KGH(s)$. These branches commence at the OL poles of the system (for $K = 0$); of these, $m$ branches terminate at OL zeros (as $K \to \infty$). The remaining $n - m$ branches follow the RL asymptotes to infinity (as $K \to \infty$) in the complex plane. The RL branches for $K < 0$ proceed from infinity and/or OL zeros toward the OL poles.
2. The RL plot is symmetric with respect to the real-axis (this is a consequence of the fact that for a real polynomial, the complex roots occur in conjugate pairs).
3. The real-axis locus, that is, the section of RL plot that aligns with the real-axis, lies to the left of an odd number of poles and zeros for $K > 0$; it lies to the right of an odd number of poles and zeros for $K < 0$.
4. The real-axis locus that lies in between a pair of poles, or zeros, contains a break point where the two RL branches split, or join together.

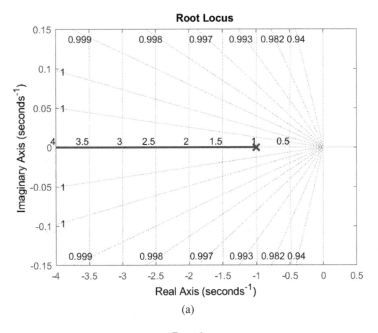

(a)

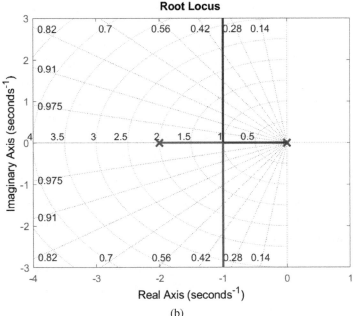

(b)

**Figure 6.2** The root locus plot for $G(s) = \frac{1}{s+1}, H(s) = 1$ (a); the root locus plot for $G(s) = \frac{1}{s(s+2)}, H(s) = 1$ (b).

The real-axis break points for both $K > 0$ and $K < 0$ are found among the solutions to the equation: $\sum \frac{1}{\sigma - p_i} - \sum \frac{1}{\sigma - z_i} = 0$.

5. The $n - m$ RL asymptotes assume angles: $\phi_a = \frac{2k+1}{n-m}(180°)$, $k = 0, 1, \ldots, n - m - 1$, (for $K > 0$), and angles: $\phi_a = \frac{2k+1}{n-m}(360°)$, $k = 0, 1, \ldots, n - m - 1$ (for $K < 0$). These asymptotes intersect at a common point on the real-axis, given as: $\sigma_a = \frac{\sum p_i - \sum z_i}{n-m}$.

The above rules suffice to sketch the RL by hand for low-order plants. Additional rules to further refine the RL plot, including the angles of departure/arrival at complex poles/zeros, can be defined (see References).

### 6.1.3 Obtaining Root Locus Plot in MATLAB

The MATLAB Control Systems Toolbox provides the "rlocus" command to plot the root locus of the loop transfer function with respect to the controller gain, $K$. The "rlocus" command takes a dynamic system object as an argument. A dynamic system object may be defined in MATLAB by any of the following commands:

(a) Use the "tf" command to define a transfer funciton model.
(b) Use the "zpk" command to define a transfer function model in factored form.
(c) Use the "ss" command to define a state variable model.
(d) Use the "frd" command to define a frequency response model.

The "grid" command adds constant damping ratio and constant natural frequency contours to the RL plot. The constant $\zeta$ lines depict the achievable damping ratios for the closed-loop roots lying on the RL branches.

The grid contours assume a prototype second-order transfer function: $T(s) = \frac{\omega_n^2}{s^2 + 2\zeta\omega_n s + \omega_n^2}$, with roots located at: $s = -\zeta\omega_n \pm \omega_n\sqrt{1 - \zeta^2}$.

The constant $\zeta$ lines are characterized by the relation: $\frac{\sigma}{\omega} = \mp\frac{\zeta}{\sqrt{1-\zeta^2}}$. In the $s$-plane, these constitute radial lines at angles defined by $\theta = \cos^{-1}\zeta$.

The following examples illustrate the sketching of the RL plot for the given loop transfer functions.

**Example 6.3:** Let $KGH(s) = \frac{K(s+3)}{s(s+2)}$; then, the characteristic polynomial is given as: $\Delta(s, K) = s^2 + (K + 2)s + 3K$. The RL plot can be sketched using the following steps:

1. The RL plot has $n = 2$ branches and $n - m = 1$ asymptote. The RL branches start at the OL poles at $0, -2$ (for $K = 0$). One of the RL

branches terminates at the zero at $s = -3$; the other follows the RL asymptote along the negative real-axis as $K \to \infty$.

2. The real-axis locus lies in the intervals: $\sigma \in (-\infty, -3] \cup [-2, 0]$ (for $K > 0$).

3. The break points are given as the solution to: $\frac{1}{\sigma} + \frac{1}{\sigma+2} = \frac{1}{\sigma+3}$ that reduces to: $\sigma^2 + 6\sigma + 6 = 0$; the equation has two solutions at: $\sigma = -1.27$ (break-away) and $-4.73$ (break-in).

4. The asymptote angle is given as: $\pm 180°$ (for $K > 0$), where $\sigma_a = 1$.

The resulting RL plot (for $K > 0$) can be sketched by hand. A computerized RL plot is generated with the help of the following MATLAB commands (Figure 6.3(a)):

```
G = zpk([-3],[0 -2],1)        % define plant transfer function
rlocus(G), grid               % plot the root locus
axis equal                    % set aspect ratio to equal
```

**Example 6.4:** Let $KGH(s) = \frac{K}{s(s+1)(s+2)}$; then, the characteristic polynomial is: $\Delta(s) = s^3 + 3s^2 + 2s + K$. The corresponding RL plot is sketched using the following steps:

1. The RL has three branches and three asymptotes. The RL branches start at the OL poles (for $K = 0$) and follow the asymptotes as $K \to \infty$.

2. The real-axis locus lies along $\sigma \in [-1, 0]$ (for $K > 0$).

3. The break points are given as the solution to: $\frac{1}{\sigma} + \frac{1}{\sigma+1} + \frac{1}{\sigma+2} = 0$, which reduces to: $3\sigma^2 + 6\sigma + 2 = 0$, and has solutions at $\sigma = -0.38, -2.62$. The first solution defines the break-away point for $K > 0$; the second one defines the break-in point for $K < 0$.

4. The asymptote angles are given as: $\pm 60°, 180°$ (for $K > 0$) and their common intersection point is $\sigma_a = -\frac{3}{3} = -1$.

The resulting RL plot (for $K > 0$) can be sketched by hand or generated with the help of following MATLAB commands (Figure 6.3(b)):

```
G = zpk([],[0 -1 -2],1)       % define plant transfer function
rlocus(G), grid               % plot the root locus
```

## 6.1.4 Stability from the Root Locus Plot

In the event when some of the RL branches cross the $j\omega$-axis, the controller gain $K$ at the stability boundary can be determined by clicking on the MATLAB plot at that location, from the polynomial stability conditions, or by using the Routh's array.

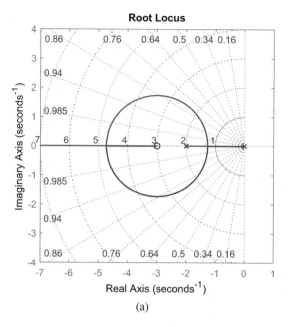

(a)

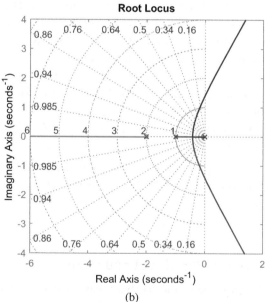

(b)

**Figure 6.3** The root locus plot for $G(s) = \frac{(s+3)}{s(s+2)}$ (a); the root locus plot for $G(s) = \frac{1}{s(s+1)(s+2)}$ (b).

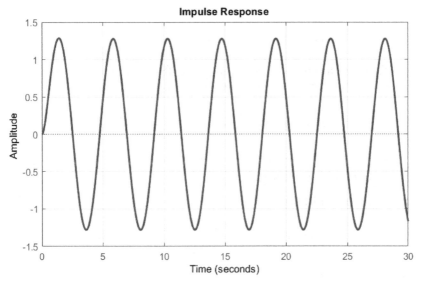

**Figure 6.4** The impulse response of the third-order system for $K = 6$.

For Example 6.4, the closed-loop characteristic polynomial is given as: $\Delta(s) = s^3 + 3s^2 + 2s + K$. The stability conditions for the third-order polynomial reduce to: $K > 0, 6 - K = 0$. Thus, the range of $K$ for stability is: $0 < K < 6$.

For $K = 6$, the closed-loop characteristic polynomial, $\Delta(s) = s^3 + 3s^2 + 2s + 6$, has roots at: $s = -3, \pm j\sqrt{2}$. Thus, when excited, the closed-loop system for $K = 6$ oscillates at a frequency of $\sqrt{2}$ rad/sec, as can be verified by using the following MATLAB commands to plot the impulse response (Figure 6.4):

```
G = zpk([],[0 -1 -2],1)        % define plant transfer function
T = feedback(6*G,1)            % closed-loop transfer function
impulse(T),grid                % plot the impulse response
```

### 6.1.5 Analytic Root Locus Conditions

The analytic RL conditions refer to the conditions that are satisfied by the points that lie on the RL plot and are the poles of the resulting closed-loop characteristic polynomial.

In order to define the analytic conditions, we assume that the loop transfer function $KGH(s)$, including any poles at the origin, is given in the factored form as:

$$KGH(s) = \frac{K(s - z_1) \ldots (s - z_m)}{(s - p_1) \ldots (s - p_n)}; \quad n > m.$$

Then, for a point $s = s_1$ to lie on the RL, and hence constitute a root of the closed-loop characteristic polynomial, it must satisfy:

$$KGH(s_1) = \frac{K(s_1 - z_1) \ldots (s_1 - z_m)}{(s_1 - p_1) \ldots (s_1 - p_n)} = -1.$$

The above relation defines two conditions, called the magnitude and the angle conditions, that are given as:

**Magnitude condition:**

$$|KGH(s_1)| = \frac{K|s_1 - z_1| \ldots |s_1 - z_m|}{|s_1 - p_1| \ldots |s_1 - p_n|} = 1.$$

**Angle condition:**

$$\angle KGH(s_1) = \angle(s_1 - z_1) + \ldots + \angle(s_1 - z_m)$$
$$- \angle(s_1 - p_1) - \ldots - \angle(s_1 - p_n) = \pm 180°.$$

From the magnitude conditions, $|KGH(s_1)| = 1$, the controller gain, $K$, to achieve a desired root location, $s_1$, on the RL is obtained as:

$$K = \frac{|s_1 - p_1| \ldots |s_1 - p_n|}{|s_1 - z_1| \ldots |s_1 - z_m|}.$$

We may note that the individual magnitude terms $|s_1 - p_i|$ and $|s_1 - z_i|$ in the above relation represent the Euclidean distances from the poles and zeros of the loop transfer function to the desired root location, $s = s_1$, in the complex plane.

Let $\theta_{z_i}, \theta_{p_i}$ denote the angles from the plant zeros and poles to the point $s_1$ on the RL plot; then, from the angle condition, the designated point on the RL plot satisfies the equalities:

$$\sum \theta_{z_i} - \sum \theta_{p_i} = \pm 180° \text{ (for } K > 0\text{)}, \quad \text{and}$$

$$\sum \theta_{z_i} - \sum \theta_{p_i} = 0° \text{ (for } K < 0\text{)}.$$

The magnitude and angle conditions are satisfied by the points lying on the RL plot. These points define the achievable closed-loop root locations when using static output feedback.

## 6.2 Static Controller Design

The static controller design involves the selection of the controller gain, $K$, from the RL plot of the loop transfer function, $KGH(s)$. The RL plot displays achievable root locations for the closed-loop characteristic polynomial, $\Delta(s) = 1 + KGH(s)$, as the controller gain $K$ varies from $0 \to \infty$.

The controller design $K$ is selected based on the design specifications defined with respect to the step response of the closed-loop system (Section 5.2). Typical unit-step response specifications include an acceptable overshoot (e.g., $OS < 10\%$ that translates as $\zeta > 0.6$) and an acceptable settling time $t_s$, where $t_s \cong \frac{4}{|\sigma|}$, $\sigma = \text{Re}[s]$.

The design specifications define a region in the complex $s$-plane where the closed-loop transfer function roots must lie. In particular, the settling time constraint dictates that the real part of the close-loop roots satisfy: $\sigma \leq -\frac{4}{t_s}$, and the damping ratio constraint requires that: $\theta \leq \pm \cos^{-1} \zeta$, where $\theta$ is the angle of the desired root location from the origin.

For example, assume that the design specifications are: $0.7 < \zeta < 1$ and $t_s < 2s$. Then, the $s$-plane region for the acceptable closed-loop root locations is bounded by: $\sigma \leq -2$ and $\theta \leq \pm 45°$ (Figure 6.5).

Assuming that a RL branch passes through a desired closed-loop root location, $s_1$, the associated controller gain $K$ can be obtained from the MATLAB generated RL plot by clicking on that location. The dialog box that

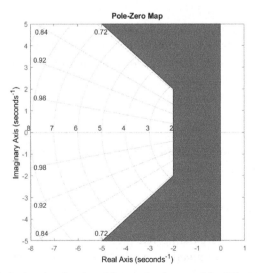

**Figure 6.5**  The desired region for closed-loop pole placement for $0.7 < \zeta < 1$ and $t_s < 2s$.

appears provides additional information on the damping, the step overshoot, and the natural frequency of the closed-loop roots for the choice of $K$ (see Figures 6.5 and 6.6). Further, the dialog box can be dragged along the RL branch to check design alternatives.

The range of K to ensure closed-loop stability can be obtained from the RL plot by clicking on a RL branch intersection with the stability boundary, that is, the $j\omega$-axis. Even though the physical plant model may be inherently stable, a high static controller gain would result in closed-loop system being unstable if the physical plant model has $n - m > 2$.

For example, in the case $n - m = 3$, the RL asymptotes are along $\theta_a = \pm 60°$, $180°$. Hence, high controller gain will inevitably lead to instability, as shown by the following example.

**Example 6.5:** Let $G(s) = \frac{1}{s(s+1)(s+2)}$, $H(s) = 1$; then, the closed-loop (CL) characteristic polynomial is given as: $\Delta(s, K) = s(s + 1)(s + 2) + K$. The polynomial is stable for $0 < K < 6$. From the RL plot (Figure 6.3(b)), the RL branches corresponding to the complex poles cross the $j\omega$-axis and enter the unstable region at $K = 6$.

Assume that we desire the closed-loop system to have a damping ratio: $\zeta = 0.7$. Then, From the RL plot (Figure 6.3(b)), we may choose a controller gain, $K = 0.65$, to choose closed-loop root locations that satisfy the damping ratio requirement. The resulting CL characteristic polynomial is factored as: $\Delta(s) = (s + 2.235)(s^2 + 0.765s + 0.291)$, and has dominant CL roots at: $-0.38 \pm 0.38$ ($\zeta = 0.7$).

Further, we can use the MATLAB Control Systems Toolbox to compute the closed-loop transfer function, plot its step response, and check that the closed-loop roots have the desired damping, by issuing the following commands:

```
G = zpk([],[0 -1 -2],1)     % define plant transfer function
T = feedback(.65*G,1)       % closed-loop transfer function
damp(T)                     % show damping and natural freq
```

The results show a damping of $\zeta = 0.7$ for the poles of the closed-loop transfer function.

## 6.3 Dynamic Controller Design

In the event that the existing RL plot does not pass through a desired root location $s_1$, as determined from the design requirements, we explore the possibility of adding a dynamic controller to the feedback loop that would suitably modify the existing RL and cause it to pass through $s_1$.

The dynamic controller $K(s)$ (aka dynamic compensator) adds poles and zeros to the loop transfer function and thus alters both the root locus plot and the characteristic polynomial.

In particular, we explore the effect of adding a first-order dynamic controller to the feedback loop, where the controller is defined by: $K(s) = \frac{K(s+z_c)}{s+p_c}$. The modified loop transfer function is given as: $K(s)G(s)H(s)$.

The first-order controller adds a zero located at $-z_c$ and a pole located at $-p_c$ to the loop transfer function. In general, addition of a finite zero to the loop transfer function causes the RL branches to bend towards it; hence, placing a zero to the left of the existing RL branches (when designing a PD or phase lead controller) adds dynamic stability to the closed-loop system by pulling the RL toward it.

The addition of a finite pole to the loop transfer function adds a new branch to the RL plot. When placed to the left of the plant poles, it repels nearby RL branches toward the imaginary axis, compromising the stability margins. When placed to the right of the plant poles, it adds a slow mode to system natural response. Placing the controller pole and zero close to the pole together diminishes the contribution of the slow mode toward system's natural response.

In the case of phase-lag or PI controllers, the controller pole–zero pair is placed close together satisfying ($p_c < z_c \ll s_1$) so as: (a) not to appreciably disturb the existing RL branches and the dominant closed-loop pole locations, (b) not to compromise the existing range of $K$ for stability; and (c) to minimize the contribution from the closed-loop pole with a large time constant to the system natural response.

In the following, we describe the transient and steady-state response improvements obtained by using first-order dynamic controllers.

### 6.3.1 Transient Response Improvement

To bring about transient response improvement in a system, a phase-lead controller may be added to the feedback loop. A similar transient response improvement can be brought by adding rate feedback or a PD controller.

**Phase-Lead Controller.** The dynamic controller used for the transient response improvement is the phase-lead controller, where the lead refers to the positive phase contributed by the controller to the Bode phase plot of the loop transfer function. The phase-lead controller is described by:

$$K(s) = \frac{K(s + z_c)}{s + p_c}; \quad z_c < p_c$$

The locations of the controller pole and zero are selected with the help of RL angle condition, as illustrated in Examples 6.6 and 6.7.

**Proportional-Derivative (PD) Controller.** In the limiting case of $p_c \to \infty$, the phase-lead controller changes into a proportional-derivative controller, described by: $K(s) = K(s + z_c)$. The location of the controller zero can be selected from the RL angle condition.

The PD controller has the undesirable effect of amplifying high-frequency noise entering the system. Hence, to achieve noise suppression, it is replaced by a phase-lead design with $p_c \gg z_c$.

**Example 6.6:** Let $G(s) = \frac{1}{s(s+2)}$; we assume that the design requirements are for the closed-loop system to have: $\zeta = 0.7$, $t_s = 2s$. We first check if a static controller can satisfy the design specifications.

For the static controller, the CL characteristic polynomial is given as: $\Delta(s) = s^2 + 2s + K$. The resulting CL roots are at: $s_{1,2} = -1 \pm \sqrt{4 - K}$. The RL plot shows two branches that break away from the real-axis at $s = -1$ (for $K = 4$) and extend to $-1 \pm j\infty$ as $K \to \infty$.

Using the static controller, the estimated settling time for the CL system response (for $K > 4$) is: $t_s = \frac{4}{|\sigma|} = 4$ s, that is, the static controller cannot meet the settling time requirement.

**PD Design.** A PD controller is defined by: $K(s) = K(s + z_c)$. In order to design the controller, let the desired root location for the closed-loop characteristic polynomial be selected as: $s_1 = -4 \pm j4$. Since the existing RL plot does not pass through this location, we must add a zero to the loop transfer function to cause the RL to bend toward $s_1$.

The zero location is selected from the RL angle condition applied to the compensated system, which states that: $\sum \theta_{z_i} - \sum \theta_{p_i} = \theta_z - 116.6° - 135° = \pm 180°$.

By applying the above angle condition (Figure 6.6), the angle subtended by the controller zero at $s_1$ is given as: $\theta_z = 71.6°$, which translates into a real-axis zero location: $z_c = -5.33$. The resulting PD controller is given as: $K(s) = K(s + 5.33)$.

With the inclusion of the PD controller, the loop transfer function is given as: $KGH(s) = \frac{K(s+5.33)}{s(s+2)}$. The closed-loop characteristic polynomial is given

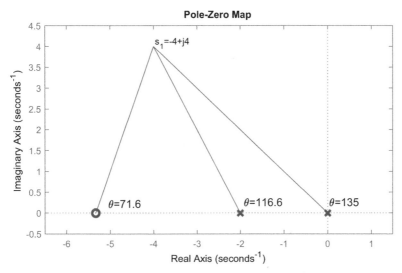

**Figure 6.6** Application of the angle criteria to choose the PD controller zero at the desired root location $s_1$: $\theta_z - \theta_{p1} - \theta_{p2} = \pm 180°$.

as: $\Delta(s) = s(s + 2) + K(s + 5.33)$. Indeed, for $K = 6$, the polynomial: $\Delta(s) = s^2 + 8s + 32$ has the dominant roots at the desired pole location $(-4 \pm j4)$ (Figure 6.7(a)).

**Phase-Lead Design.** The phase-lead controller adds a pole–zero pair to the loop transfer function, where, from the angle condition, $\theta_z - \theta_p = 71.6°$. We may arbitrarily choose a location for either $z_c$ or $p_c$ and use this constraint to determine the other.

For the above example, we may choose: $z_c = -5$, $p_c = -30$ to obtain the following phase-lead controller: $K(s) = \frac{K(s+5)}{s+30}$.

The resulting closed-loop characteristic polynomial is given as: $\Delta(s) = s^3 + 32s^2 + (60 + K)s + 5K$. Then, for $K = 200$, the equation has dominant closed-loop roots at $-4.8 \pm j4.6$ ($\zeta = 0.72$) that meet the design specifications (Figure 6.7(b)).

The MATLAB commands for the above PD and phase-lead controller designs are given below:

```
G=zpk([],[0 -2],1)          % define plant transfer function
Kpd=zpk(-5.33,[],1)         % define PD controller
Klead=zpk(-5,-30,1)         % define phase-lead controller
rlocus(Kpd*G), grid         % root locus with PD controller
rlocus(Klead*G), grid       % RL with phase-lead controller
```

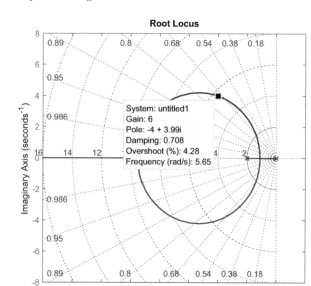

(a)

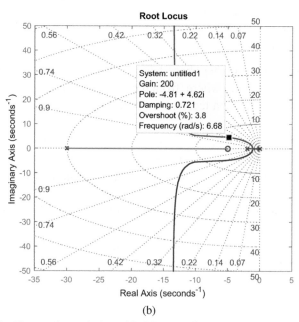

(b)

**Figure 6.7**   The root locus design with: PD controller (a); phase-lead controller (b).

**Example 6.7:** Let $G(s) = \frac{1}{s(s+2)(s+5)}$; we assume that the design specifications for a response to unit-step input are given as: $\%OS \leq 10\%$, $t_s \leq 2s$.

An RL plot of the loop transfer function: $KG(s)$ shows that for $K = 8.8$, dominant closed-loop roots located at: $= -0.76 \pm j1.01$ ($\zeta = 0.6, OS \cong 10\%$), with a settling time of $t_s = 5.3$ s. Thus, the static gain controller is inadequate in meeting the settling time requirement.

**Phase-Lead Design.** In order to meet the settling time requirement, let the desired dominant CL root locations be selected as: $s = -2 \pm j2.5$ ($\zeta = 0.62$). We consider a phase-lead controller of the form: $K(s) = \frac{K(s+z_c)}{s+p_c}$, and use the angle condition to determine the phase contribution from the controller as: $\theta_z - \theta_p = -180° + 129° + 90° + 40° \cong 79°$.

We arbitrarily choose a controller zero location: $z = 2$, and use the angle condition to obtain: $p \cong 15$, that is, $K(s) = \frac{K(s+2)}{s+15}$. The RL plot of $K(s)G(s)$ (Figure 6.8(a)) has, for $K = 150$, dominant closed-loop roots located at $s = -2.06 \pm j2.28$ ($\zeta = 0.67$).

The resulting closed-loop characteristic polynomial: $s^3 + 20s^2 + 75s + 165$, has a third root located at $s = -15.87$.

The MATLAB commands for the above phase-lead design are given below:

```
G=zpk([],[0 -2 -5],1)        % define plant transfer function
rlocus(G), grid, hold        % plot the root locus
Klead=zpk(-2,-15,1)          % define phase-lead controller
rlocus(Klead*G), hold        % RL with phase-lead controller
T=feedback(150*Klead*G,1);   % closed-loop transfer function
pole(T)                      % closed-loop pole locations
```

In the above example, in order to meet the angle constraint for dominant complex poles with the required $\zeta$, the controller zero should be placed at or to the left of the plant pole at $s = -2$. In fact, the controller zero may be selected in the interval: $[-2.4, -2]$.

We note that the selected controller zero location, $z_c = -2$, coincides with a plant pole, the loop transfer function involves a pole–zero cancellation and simplifies to: $KGH(s) = \frac{K}{s(s+5)(s+15)}$. Such pole–zero cancellations are permissible as long as they are restricted to the open left-half plane (OLHP) excluding the imaginary axis.

A pole–zero cancellation in the model-based controller design, however, does not imply the same in practice. In fact, modeling inaccuracies in the plant and the component tolerances in realizing the controller make the exact pole–zero cancellation an unlikely event in practice.

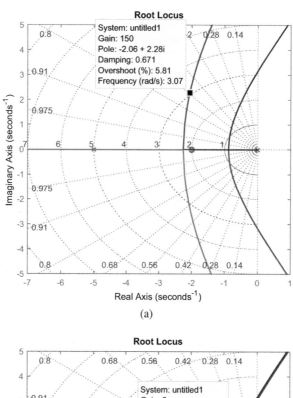

(a)

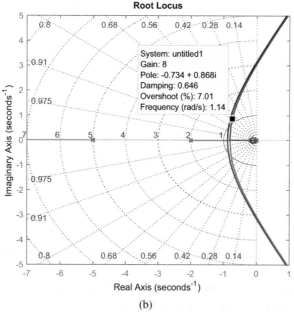

(b)

**Figure 6.8**   RL improvement with phase lead design (a); steady-state velocity error constant improvement with phase-lag design (b).

## 6.3.2 Steady-State Error Improvement

The steady-state tracking error characterizes the long-term behavior of the closed-loop control system when a constant step reference input or a linearly varying ramp reference input is applied (Section 5.3). For a given plant described by its transfer function, $G(s)$, the steady-state tracking error to a step or a ramp input is computed using the position and velocity error constants. These constants are defined in terms of the low-frequency loop gain as:

$$K_p = \lim_{s \to 0} KGH(s)$$
$$K_v = \lim_{s \to 0} sKGH(s).$$

In the case of unity-gain feedback configuration ($H(s) = 1$), the steady-state tracking errors to step and ramp inputs are computed as:

$$e(\infty)|_{step} = \frac{1}{1 + K_p}$$
$$e(\infty)|_{ramp} = \frac{1}{K_v}.$$

The controller design for steady-state error improvement aims to boost $K_p$ and/or $K_v$ by adding a PI or a phase-lag controller to the feedback loop. Both controllers add a pole–zero pair to the loop transfer function and are described by:

$$K(s) = \frac{K(s + z_c)}{(s + p_c)}, \quad p_c < z_c,$$

where $p_c = 0$ denotes the PI controller.

With the addition of the phase-lag or PI controller, the position error constant is modified as: $K'_p = (z_c/p_c)K_p > K_p$.

The PI controller, in particular, adds an integrator to the loop; hence, $K'_p = \infty$, which results in $e(\infty)|_{step} = 0$.

The pole–zero locations selected for the phase-lag or PI controller design must obey: $p_c < z_c \ll s_1$, where $s_1$ denotes the dominant closed-loop root location for an acceptable transient response. As a result, the angle contributed by the controller pole–zero pair stays small, that is, $\theta_z - \theta_p \approx 0°$. Hence, the controller does not to appreciably affect the existing RL plot that is assumed to pass through the desired pole location $s_1$.

On the flip side, the closed-loop pole added by the phase-lag or PI controller is located close to the origin, which introduces a natural response mode with a large time constant. Nevertheless, the contribution of this mode to the overall system response remains small due to the close proximity of the closed-loop pole to the controller zero.

**Example 6.8:** Let $G(s) = \frac{1}{s(s+2)(s+5)}$; we assume that the design specifications are: $\zeta \geq 0.6, e_{ss}|_{ramp} \leq 0.1$.

We first note that the transient response requirement can be met with a static controller ($K = 8$), which results in the closed-loop roots at: $s = -0.78 \pm j92$ ($\zeta = 0.65$).

Next, using $KG(s) = \frac{K}{s(s+2)(s+5)}$, we obtain a velocity error constant, $K_v = K/10$, and for the selected design ($K = 8$), we obtain: $K_v = 0.8$.

Since we desire to increase $K_v$ to 10, we may consider the a phase-lag controller: $K(s) = \frac{K(s+.1)}{s+0.008}$, so that the compensated system has a velocity error constant: $K_v = \lim_{s \to 0} sK(s)G(s) = 1.25$ K. Then, for $K = 8$, we have $K_v = 10$, with $e_{ss}|_{ramp} \cong 0.1$.

The resulting CL roots are given as: $-0.11, -5.43, -0.79 \pm j0.93$ ($\zeta \cong 0.65$); the dominant poles have a settling time of $t_s = 5.6$s.

The MATLAB commands for the phase-lag design are given below (Figure 6.8(b)):

```
G=zpk([],[0 -2 -5],1)          % define plant transfer function
rlocus(G), grid, hold          % plot the root locus
Klag=zpk(-.1,-.008,1)          % define phase-lag controller
rlocus(Klag*G), hold           % RL with phase-lag controller
```

### 6.3.3 Lead–Lag and PID Designs

When the controller design specifications require simultaneous improvements to transient response and steady-state tracking error, a lead-lag or a PID controller may be considered. A lead-lag controller represent a cascade of a phase-lead stage and a phase-lag stage. A PID controller similarly combines PD and PI controllers.

The design steps for a lead-lag/PID controller design are given as follows:

1. Obtain a desired dominant pole location ($s_i$) from the transient response specifications.
2. Check the RL plot for the loop transfer function to determine if a static controller suffices.
3. If necessary, design a phase-lead or PD controller for the system. Plot the RL for the compensated system and verify that it passes through the desired location ($s_i$). Determine the RL gain at ($s_i$).

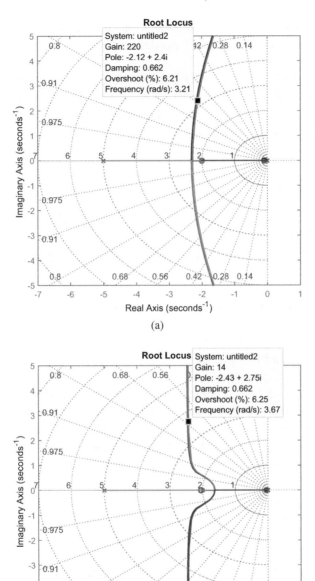

**Figure 6.9** Root locus improvement with lead–lag compensation (a); RL improvement with PID compensation (b).

4. Evaluate the relevant error constant to determine if steady-state error specifications are met. If needed, design a phase-lag or PI section to boost the error constant.
5. Plot the RL for $K(s)G(s)$, where $K(s) = K_{lead}(s)K_{lag}(s)$, or equivalently, $K(s) = K_{PD}(s)K_{PI}(s)$, and verify that it passes through the desired location $(s_i)$. Also check the RL gain, $K$, and verify that error constant requirements are met.

**Example 6.9:** Let $G(s) = \frac{1}{s(s+2)(s+5)}$; we assume that the design specifications are: $\%OS \leq 10\%$, $t_s \leq 2\text{s}$, $e_{ss}|_{ramp} \leq 0.1$. These translate into: $\zeta \geq 0.6$, $|\sigma| \geq 2$, $K_v \geq 10$.

**Lead-Lag Design.** The steps for a lead-lag design are given as follows (Figure 6.9(a)):

1. A RL plot of $KG(s)$ shows dominant closed-loop roots at $s = -0.76 \pm j1.01 (\zeta = 0.6)$ with $t_s = 5.3$ s. Thus, a static gain controller does not suffice.
2. Assume that the desired closed-loop pole locations are: $s = -2.2 \pm j2.5$ $(\zeta = 0.65)$. Use the angle condition to obtain: $\theta_z - \theta_p = -180° + 131° + 95° + 42° = 88°$.
3. **Phase-lead Design:** let $z_c = 2, p_c = 20, K_{lead}(s) = \frac{K(s+2)}{s+20}$.
   The RL plot of $K_{lead}(s)G(s)$ has, for $K = 220$, dominant CL roots at: $s = -2.15 \pm j2.49 (\zeta = 0.65)$. We may note that this is a "ballpark" design that approximately satisfies the angle condition.
4. **Phase-lag Design:** For $K_{lead}(s)G(s)$ with $K = 220$, we obtain $K_v = 2.25$, that is, we need to improve $K_v$ by a factor of 4.5; accordingly, let $z_c = 0.09, p_c = 0.02, K_{lag}(s) = \frac{K(s+0.09)}{s+0.02}$.
5. The RL plot of $K_{lead}(s)K_{lag}(s)G(s)$ has, for $K = 225$, dominant CL roots at $s = -2.12 \pm j2.46 (\zeta = 0.65)$. The resulting $e_{ss}|_{ramp} \leq 0.099 (K_v = 10.13)$.
6. The final lead-lag controller is given as: $K(s) = K(s+0.09)(s+2)/(s+0.02)(s+20), K = 225$.

The MATLAB commands for the above lead-lag are given below (Figure 6.9(b)):

```
G = tf(1, [1 7 10 0]);          % define plant transfer function
Klead = tf([1 2],[1 20]);       % define phase-lead controller
rlocus(G*Klead), grid, hold     % RL with phase-lead controller
Klag = tf([1 .09],[1 .02]);     % define phase-lag controller
rlocus(G*Klead*Klag), hold      % RL with Lead-lag controller
```

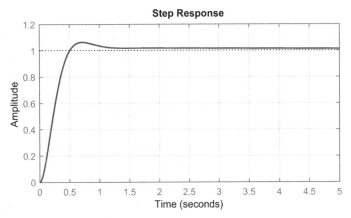

**Figure 6.10** Step response of phase-lead plus PI compensated system (Example 6.10).

**PID Design.** Alternatively, the steps for a PID design are given as follows (Figure 6.9(b)):

1. **PD design:** let $z_c = 2.05$; $K_{PD}(s) = K(s + 2.05)$.
2. The RL plot of $K_{PD}(s)G(s)$ has, for $K = 12.5$, dominant CL roots at $s = -2.4 \pm j2.5$ ($\zeta = 0.69$). For the PD compensated system, $K_v = 1.25$.
3. **PI design:** let $z_c = 0.05$, $p_c = 0$, $K_{PI}(s) = K(s + 0.05)/s$.
4. The RL plot of $K_{PD}(s)K_{PI}(s)G(s)$ has, for $K = 12.5$, dominant CL roots at $s = -2.4 \pm j2.5$($\zeta = 0.69$). The resulting $e_{ss}|_{ramp} = 0$ ($K_v = \infty$).
5. The PID controller is given as: $K(s + 0.05)(s + 2.05)/s$, $K = 12.5$.

The MATLAB commands for the PID design are given below (Figure 6.9(b)):

```
G=tf(1, [1 7 10 0]);        % define plant transfer function
Kpd = tf([1 2.05],1);       % define PD controller
rlocus(G*Kpd)               % root locus with PD controller
Kpi = tf([1 .05],[1 0]);    % define PI controller
rlocus(G*Kpd*Kpi)           % root locus with PID controller
```

When designing controllers for simultaneous transient and steady-state response improvement, we can also combine the phase-lead and PI designs, as shown by the following example.

**Example 6.10:** Let $G(s) = \frac{1}{s(s+2)}$; we assume that the design requirements are for the closed-loop system to have: $\%OS \leq 10\%$, $t_s \leq 2$s, $e_{ss}|_{ramp} \leq 0.1$.

The design steps are given as follows:

1. A RL plot of $KG(s)$ shows that static gain controller is inadequate.
2. **Phase-lead Design.** Assume that the desired closed-loop pole location is given as: $s = -4.5 \pm j4.5$ ($\zeta = 0.7$); then, the desired phase-lead controller is given as: $K_{lead}(s) = \frac{K(s+2)}{s+9}$ (Example 6.6). Further, for $K = 40$, we obtain the closed-loop poles at $s = -4.5 \pm j4.45$ ($\zeta = 0.71$).
3. **PI Design.** For the phase-lead compensated system, $K_{lead}(s)G(s)$ with $K = 40$, we obtain $K_v = 4.44$. In order to improve the $K_v$, we propose a PI controller, where $K_{PI}(s) = \frac{K(s+0.1)}{s}$.
4. The overall controller design is given as: $K(s) = \frac{K(s+0.1)(s+2)}{s(s+9)}$, $K = 40$. The resulting closed-loop characteristic polynomial is: $\Delta(s) = s^3 + 9s^2 + 40s + 4$, which has roots at: $s = -4.45 \pm j4.4, -0.102$.

   The step response of the CL system shows that the response settles to within 2% of the final value in about 1.2 s and to within 1% in about 4.5 s (Figure 6.10).

We note that both phase-lag and PI designs result in one of the closed-loop roots located close to the origin at $\sigma = -0.102$. This root introduces a natural response mode with a long time constant of $\tau = 10$ s. However, the contribution of this mode toward system step response is relatively minor due to the close proximity of the CL transfer function zero.

Using partial fractions followed by inverse Laplace transform, the output of the compensated system to a unit-step input is given as:

$$y(t) = 1.0003 + 0.0235e^{-0.102t} - 1.0238e^{-4.45t}$$
$$(\cos 4.393t + 1.0123\sin 4.393t)$$

Hence, the slow response mode with $\tau = 10s$ contributes only about 2% to the unit-step response. The following MATLAB commands are used to check the performance of the controller:

```
G=zpk([],[0 -2],1)            % define plant transfer function
K=zpk([-2 -.1],[0 -9],1)      % define phase-lead controller
T=feedback(40*G*K,1)          % closed-loop transfer function
step(T)                       % plot the step response
axis([0 5 0 1.2])             % set axis limits
```

### 6.3.4 Rate Feedback Compensation

A rate feedback controller (Section 4.2, Figure 4.4), similar to a PD controller, improves relative stability and transient response of the closed-loop system.

In order to analyze the rate feedback control system, let the plant transfer function be given as: $G(s) = \frac{n(s)}{d(s)}$; then, the closed-loop transfer function for the rate feedback configuration is obtained as:

$$\frac{y(s)}{r(s)} = \frac{Kn(s)}{d(s) + (k_f s + K)n(s)}.$$

The resulting closed-loop characteristic polynomial is given as: $\Delta(s) = d(s) + (k_f s + K)n(s)$. The closed-loop system poles can be placed at desired locations through the choice of $K$ and $k_f$. The use of rate feedback offers additional flexibility in controller design, as shown by the following example.

**Example 6.11:** Let $G(s) = \frac{1}{s(s+2)}$; we assume that the design requirements are for the step response of the closed-loop system to have: $\%OS \leq 10\%$, $t_s \leq 2$s.

The desired CL pole locations are selected as: $s = -4.5 \pm j4.5 \, (\zeta = 0.7)$. These locations corresponds to a desired closed-loop characteristic polynomial: $\Delta_{des}(s) = s^2 + 9s + 40.5$.

Then, by comparing the polynomial coefficients with those of the rate feedback characteristic polynomial: $\Delta(s) = s^2 + (k_f + 2)s + K$, we determine the controller gains as: $k_f = 7$, $K = 40.5$.

The RL technique can be used to accomplish the design of rate feedback controller in two steps: in the first step, the inner (minor) loop is designed for desired closed-mirror-loop pole locations; then, in the second step, those locations serve as open-loop pole locations for the outer loop design. The two-step design is explained below.

**Minor Loop Design.** The loop transfer function for the minor loop design is given as: $G(s)k_f s$. Let $G(s) = \frac{n(s)}{d(s)}$; then, the resulting closed-minor-loop transfer function is obtained as: $G_{ml}(s) = \frac{n(s)}{d(s) + n(s)k_f s}$.

**Outer Loop Design.** The loop transfer function for the outer loop design is given as: $KG_{ml}(s)$. The resulting closed-loop transfer function is given as: $T(s) = \frac{n(s)}{d(s) + n(s)(k_f s + K)}$.

**Example 6.12:** Let $G(s) = \frac{1}{s(s+2)}$; we assume that the design requirements are for the unit-step response to achieve: $\%OS \leq 10\%, t_s \leq 2$s.

**Minor Loop Design.** The minor loop gain is: $\frac{k_f s}{s(s+2)}$, so that the closed-minor loop transfer function is: $G_{ml}(s) = \frac{1}{s(s+2+k_f)}$. By choosing, for example, $k_f = 2$, we obtain: $G_{ml}(s) = \frac{1}{s(s+4)}$. The gain selection is performed on the root locus plot (Figure 6.11(a)).

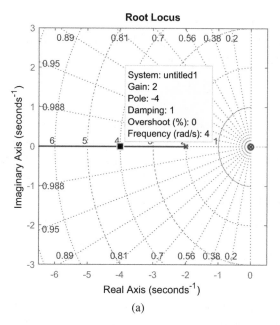

(a)

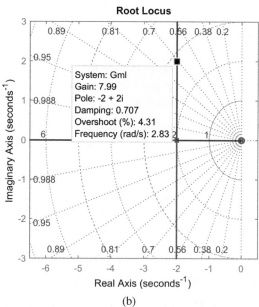

(b)

**Figure 6.11** Rate feedback design for Example 6.12: minor loop design (a); outer loop design (b).

**Outer Loop Design.** From the RL plot for $KG_{ml}(s)$ in Figure 6.11(b), we may choose, for example, $K = 8$, for the overall closed-loop roots located at: $s = -2 \pm j2$.

The MATLAB script for the two stage design is given below:

```
G=tf(1, [1 2 0]);           % define plant transfer function
s=tf('s');                  % rate feedback transfer function
rlocus(G*s), grid, hold     % root locus plot for loop gain
axis([-6.5 .5 -3 3])        % set axis limits
Gml=feedback(G,2*s)         % minor loop transfer function
rlocus(Gml), hold           % root locus for outer loop design
```

We may note that although the minor loop transfer function, $G(s)K_f s$, includes a zero at the origin, this zero does not cancel any existing plant poles at that location. In other words, closing the inner feedback loop through the rate (derivative) term does not affect the system type, which is defined as the number of open-loop poles at the origin.

In the above example, the plant transfer function is of type 1; this categorization also holds for the closed-minor-loop transfer function, $G_{ml}(s)$.

**Example 6.13:** Let $G(s) = \frac{1}{s(s+2)(s+5)}$; we assume that the design specifications for the step response are given as: $\%OS \le 10\%, t_s \le 2$ s; these specifications translate into $\zeta \ge 0.6, |\sigma| \ge 2$.

We consider the rate feedback design using $H(s) = K_f s$. The feedback gain $K_f$ is selected on the minor loop RL plot for $K_f s \, G(s) = \frac{K_f s}{s(s+2)(s+5)}$.

**Minor Loop Design.** The minor loop RL (Figure 6.12(a)) contains two branches that split at $\sigma = -3.5$ ($K_f = 2.25$) and follow the asymptotes at $\pm 90°$.

We may, for example, choose $K_f = 15$; the resulting closed-minor-loop roots are located at: $s = -3.5 \pm j3.57$ ($\zeta \cong 0.7$). The loop transfer function for the outer loop design is given as:

$$KG_{ml}(s) = \frac{K}{s[(s+2)(s+5) + K_f]} = \frac{K}{s[s^2 + 7s + 25]}.$$

**Outer Loop Design.** From the outer loop RL (Figure 6.12(b)), we may choose $K = 34$ for the resulting closed-loop roots at: $s = -2.27 \pm j2.95$ ($\zeta = 0.61$), with $t_s \cong 1.8$ s.

The MATLAB commands for the rate feedback design are given below (Figure 6.12):

```
G=tf(1, [1 7 10 0]);        % define plant transfer function
s=tf('s');                  % rate feedback transfer function
rlocus(G*s), grid, hold     % root locus plot for loop gain
```

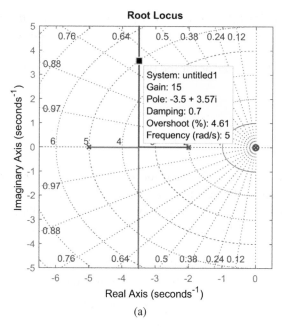

(a)

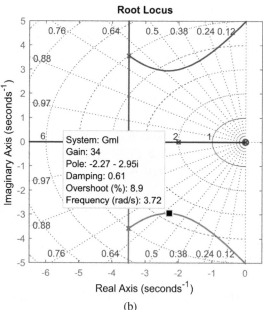

(b)

**Figure 6.12**   Rate feedback design for Example 6.13: minor loop design (a); outer loop design (b).

```
axis([-6 1 -6 6])                    % set axis limits
Gml=feedback(G,15*s)                 % minor loop transfer function
rlocus(Gml), hold                    % root locus for outer loop design
```

**Analytical Rate Feedback Design.** We note that the closed-loop characteristic polynomial for the rate feedback configuration (Figure 4.4) is given as: $\Delta(s) = s[(s+2)(s+5) + k_f] + K$. Hence, we can analytically choose the controller gains, $K$ and $K_f$, to achieve the desired closed-loop roots for $\Delta(s)$. For example, choosing $K_f = 15$, $K = 35$, realizes the above design with the resulting closed-loop roots at: $s = -2.27 \pm j2.95 (\zeta = 0.61)$.

## 6.3.5 Controller Designs Compared

The feedback controller design choices presented in this chapter include: static gain controller, phase-lag, phase-lead, lead-lag, PID, and the rate feedback controller. Here, we compare the performance of these controllers (Examples 6.5, 6.7, 6.8, 6.9 and 6.13) for the plant model: $G(s) = \frac{1}{s(s+2)(s+5)}$. The controllers were designed to achieve a damping ratio $\zeta \cong 0.6$ for the dominant closed-loop roots. The controller designs are summarized in Table 6.1.

The performance of the controllers is compared by plotting the step response of the compensated system (Figure 6.13). The comparison includes the overshoot, settling time, and steady-state tracking error for the six controller types (Table 6.1). We make the following observations from the table:

**Table 6.1** A comparison of the unit step response of the compensated system for six different compensator designs for $G(s) = \frac{1}{s(s+2)(s+5)}$

| Controller Type | Controller Transfer Function, $K(s)$ | Dominant Closed-loop Pole Locations | Rise Time | Percentage Overshoot | Settling Time |
|---|---|---|---|---|---|
| Static | 8.75 | $-0.77 \pm j1.0$ | 2.46s | 9% | 4.9s |
| Phase-lag | $\dfrac{8.75(s+.1)}{s+.008}$ | $-0.72 \pm j0.95$ | 2.17s | 20% | 17.4s |
| Phase-lead | $\dfrac{220(s+2)}{s+20}$ | $-2.16 \pm j2.44$ | 0.99s | 6% | 1.9s |
| Lead-lag | $\dfrac{220(s+2)(s+.1)}{(s+20)(s+.008)}$ | $-2.11 \pm j2.39$ | 0.93s | 10% | 8.1s |
| PID | $\dfrac{16.75(s^2 + 2.1s + .205)}{s}$ | $-2.43 \pm j3.22$ | 0.66s | 11% | 1.6s |
| Rate feedback | $15s + 35$ | $-2.21 \pm j2.95$ | 1.55s | 1% | 1.4s |

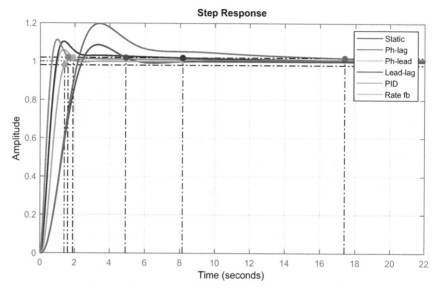

**Figure 6.13**   A comparison of the step responses for the various compensator designs.

1. The static gain and phase-lag controllers do not provide the desired transient response improvement; hence, their step responses have long settling times.
2. The PID controller has the fastest rise time followed by the lead-lag and phase-lead designs.
3. The rate feedback controller offers the shortest settling time followed by PID and phase-lead design.
4. The phase-lag controller has highest overshoot (20%); the rate feedback controller has the lowest overshoot (1%).
5. The rate feedback controller offers the best overall performance. The results, however, cannot be generalized.

**Controller Design Using MATLAB SISO Tool.** The MATLAB Control System Toolbox provides a utility for controller design in the case of single-input single-output (SISO) systems. The design tool is accessed by typing "sisotool" on the command line interface or by entering the Control System Designer App. For example, we may initialize the design tool as:

```
sisotool(tf(1,[1 2 0]))   % start MATLAB SISO design tool
```

Subsequently, we can add controller poles and zeros under the controller editor tab and observe the step response under the analysis plots tab. The interested readers may refer to the MATLAB Control Systems Toolbox documentation for further details.

## 6.4 Controller Realization

The dynamic controlles of the phase-lead, phase-lag or PD, PI, PID types represent frequency selective filters that may be realized by electronic circuits built with operational amplifiers, resistors and capacitors. The general procedure for realizing a given transfer function is described below.

**Operational Amplifier Circuits.** An operational amplifier in the inverting configuration (Figure 6.14) has the input–output transfer function:

$$\frac{V_0(s)}{V_i(s)} = -\frac{Z_f(s)}{Z_i(s)}$$

where $Z_i(s)$ and $Z_f(s)$ denote the input and feedback path impedances. These impedances comprise $RC$ networks connected in series or in parallel with their impedances computed as:

$$\text{Series RC circuit: } Z_{ser}(s) = R + \frac{1}{Cs} = \frac{RCs + 1}{Cs}$$

$$\text{Parallel RC circuit: } Z_{par}(s) = \frac{R/Cs}{R + \frac{1}{Cs}} = \frac{R}{RCs + 1}$$

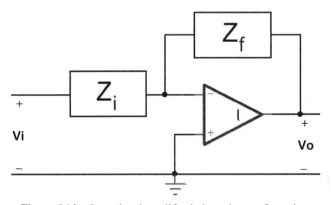

**Figure 6.14**  Operational amplifier in inverting configuration.

### 6.4.1 Phase-Lead/Phase-Lag Controllers

A phase-lead or phase-lag controller is realized by selecting parallel RC circuits in the input and feedback paths of an operational amplifier. The resulting controller transfer function is given as:

$$K(s) = -\frac{Z_f(s)}{Z_i(s)} = -\frac{R_f}{R_i}\frac{(R_iC_is+1)}{(R_fC_fs+1)}.$$

The controller transfer function places a zero at: $z_c = \frac{1}{R_iC_i}$, and a pole at: $p_c = \frac{1}{R_fC_f}$. Hence, we may choose $R_iC_i > R_fC_f$ for the phase-lead and $R_iC_i < R_fC_f$ for the phase-lag design. For the sign correction, a resistive op-amp circuit with $R_i = R_f$, static gain: $\frac{V_0}{V_i} = -\frac{R_f}{R_i} = -1$ can be employed.

**Example 6.11:** Let $K(s) = \frac{5(s+1)}{s+10}$; then, we have $R_iC_i = 1$, $R_fC_f = 0.1$, $R_f/R_i = 0.5$. We may choose, for example, $R_i = 100$ K$\Omega$. Then, $R_f = 50$ K $\Omega$, $C_i = 10\,\mu$ F, $C_f = 2\,\mu$ F.

### 6.4.2 PD, PI, PID Controllers

These controllers may be realized by combining the following impedances:

$$G_{PD}(s) = -\frac{R_f}{Z_{par}(s)} = -\frac{R_f}{R_i}(R_iC_is+1)$$

$$G_{PI}(s) = -\frac{1/C_fs}{Z_{par}(s)} = -\frac{1}{R_iC_f}\frac{(R_iC_is+1)}{s}$$

$$G_{PID}(s) = -\frac{Z_{f-ser}(s)}{Z_{i-par}(s)} = -\frac{1}{R_iC_f}\frac{(R_iC_is+1)(R_fC_fs+1)}{s}.$$

Further, in the case of the PID controller, the PID gains can be solved as functions of component values as:

$$k_p = \frac{R_f}{R_i} + \frac{C_i}{C_f}, \quad k_i = R_fC_i, \quad k_d = \frac{1}{R_iC_f}.$$

**Example 6.12:** Let $G_{PID}(s) = \frac{(s+0.1)(s+10)}{s}$; then, to realize it with RC networks, we may choose, for example: $R_i = 100$ K$\Omega$, $R_f = 1$ M$\Omega$, $C_i = 1\,\mu$ F, $C_f = 10\,\mu$ F.

## Skill Assessment Questions

Link to the answers:
http://www.riverpublishers.com/book_details.php?book_id=449

1. Consider the model of a DC motor with shaft angular position output, given as: $G(s) = \frac{5000}{s(s+10)(s+100)}$. Assume that the motor is connected in unity-gain feedback configuration.

   (a) Sketch the root locus (RL) for the motor model.
   (b) Choose a controller gain $K$ to achieve $\zeta \cong 0.7$.
   (c) Find the range of $K$ for stability from the RL plot.

2. Let $G(s) = \frac{s+3}{s(s+1)(s+2)}$; assume that the system is connected in unity-gain feedback configuration.

   (a) Plot the root locus for the system and find the range of $K$ for stability.
   (b) Design a phase-lead controller for unity gain feedback to meet the following specifications: $\omega_n \cong 2\sqrt{2}, \zeta \cong 0.7$.

3. Consider the plant in Question 6.2 with the phase-lead controller.

   (a) Add a phase-lag controller to boost the velocity error constant to $K_v > 10$.
   (b) Plot the ramp response to verify $e(\infty)|_{ramp} < 0.1$.
   (c) Realize the controller using op-amp circuits.

4. Consider the simplified model of a flexible structure given as: $G(s) = \frac{0.5}{s} + \frac{0.1}{s^2+s+100}$. Assume that the structure is connected in unity-gain feedback configuration.

   (a) Design a PID controller for the model (use the MATLAB "pidtune" command)
   (b) Realize the controller using op-amp circuits.

5. Consider the model of the DC motor in Question 6.1.

   (a) Design a rate feedback controller to achieve: $\omega_n \geq 50\frac{rad}{s}, \zeta \geq 0.6$.
   (b) Plot the step response of the closed-loop system.

6. Consider the model of human postural dynamics in the sagittal plane described as a rigid inverted pendulum, given as: $G(s) = \frac{k}{s^2-\Omega^2}$, where $k = 0.01, \Omega = 3\,rad/sec$.

   (a) Design a Phase-lead controller to stabilize the model and achieve the following specifications: $\omega_n \geq 5\frac{rad}{s}, \xi \geq 0.8$.

(b) Add a phase-lag controller to achieve less than 2% steady-state error to a step reference input.

7. Consider the input-output model of room heating given by: $G(s) = \frac{K}{\tau s+1}$, where $K = 0.2$, $\tau = 20$.

   (a) Design a PID controller for the model to achieve a settling time of less than 10 sec with less than 5% overshoot.

   (b) Plot the step response of the closed-loop system.

8. The model of the pitch axis of a quad-copter with actuator is described as: $G(s) = \frac{1}{s^2(s+10)}$. It is desired to have the command response settle in 3 sec.

   (a) Design a Phase-lead controller to achieve the desired settling time.

   (b) Plot the step response of the closed-loop system.

9. Consider the input-output model of magnetic levitation of a small metallic ball, given as: $G(s) = \frac{5K}{s^2+20}$, where $K = 100$.

   (a) Design a PID controller for the model to achieve a settling time of less than 1 sec with less than 20% overshoot.

   (b) Plot the step response of the closed-loop system.

10. A servomechanism for position control is described by the plant transfer function: $G(s) = \frac{10}{s(s+1)(s+10)}$. The system is desired to have a settling time of 0.5 sec and less than 5% error to a ramp input.

   (a) Design a phase-lead controller to achieve transient response specifications.

   (b) Add a phase-lag controller to achieve the steady-state error specifications.

   (c) Use op-amp circuits to realize the controller.

# 7

# Design of Sampled-Data Systems

## Learning Objectives

1. Obtain pulse transfer function of a sampled-data system model.
2. Obtain unit-pulse and unit-step responses of sampled-data systems.
3. Analyze the stability of the closed-loop sampled data system models.
4. Design and emulate controllers for sampled-data system models.

The aim of the control system design is to improve the behavior of a physical plant through the use of a controller that conditions an error signal to generate the plant input signal. The controller designed using root locus or other suitable techniques can be implemented using analog circuits or, alternatively, in computer code using digital signal processing tools. The measurement sensor may similarly generate sampled output for feedback. This chapter explores controller design for implementation with clock-driven devices, that is, microcontrollers and computers.

In contemporary control systems technology, data acquisition card (DAQ) is used to sense, sample, and process plant output and other variables of interest, and the process controller is implemented as a software routine on a programmable logic controller (PLC), microcontroller, or digital signal processor (DSP). The DAQ includes an analog-to-digital converter (ADC) that samples, quantizes, and temporarily stores the variables. The output of the DAQ is a sequence of numbers that represent quantities at multiples of sampling period $T$.

The control systems that involve sampled variables and software-based controllers are analyzed using techniques designed for the sampled-data systems. These discrete-time systems are modeled using difference equations that operate on number sequences. Besides a physical plant and a digital controller, such systems include sample and hold elements representing data acquisition, and analog-to-digital and digital-to-analog conversion (Figure 7.1).

167

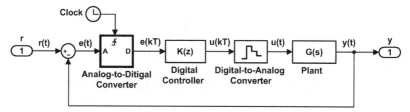

**Figure 7.1**   A closed-loop sampled-data system model.

The output of the digital controller is a discrete-time sequence: $u(kT)$, $k = 0, 1, \ldots$, that represents the sampled values of a continuous signal $u(t)$. Before applying it to the process, the controller output is converted to a continuous-time signal using a hold device, such as a zero-order hold (ZOH), that is, a device that holds its output constant for one time period.

The choice of the sampling period is an important consideration in digital control system design. In general, the sampling frequency should be selected 4–10 times higher than the system bandwidth, approximated as the inverse of the smallest time constant of the system. A lower sampling frequency causes phase margin degradation, adversely affecting relative stability of the closed-loop system.

The root locus design of the digital controller is based on the pulse transfer function of the plant that is obtained using $z$-transform and describes its input–output behavior at the sampling intervals. The rules for plotting the root locus for the discrete-time systems in the complex $z$-plane are the same as those for the continuous-time systems in the $s$-plane. The stability of the closed-loop sampled-data system in the $z$-plane, however, requires that the desired characteristic polynomial has its roots confined to the unit disk. The controller design can be performed for desired settling time and damping ratio of the closed-loop roots.

Controller emulation is a practical and time-saving approach to obtain an approximate digital controller that emulates an already designed analog controller for the continuous-time plant. Assuming a high enough sampling rate, the digital controller obtained by emulation gives comparable performance to the analog controller it mimics.

In this chapter, we will discuss models of sampled-data system, their properties, stability characterization, time response, and the analysis and controller design for such systems.

## 7.1 Models of Sampled-Data Systems

The sampled-data systems operate on discrete-time, that is, the variables change values at integral multiples of the sampling period. To model the sampled-data systems, we consider an ideal sampler that samples a physical signal $r(t)$ every $T$ seconds and generates a series of impulses with weights $r(kT), k = 0, 1, \ldots$. Mathematically, the sampler output, $r^*(t)$, represents multiplication of $r(t)$ with a train of impulses and is given as:

$$r^*(t) = \sum_{k=0}^{\infty} r(kT)\delta(t - kT).$$

We apply the Laplace transform to the sampled signal $r^*(t)$ to obtain:

$$r^*(s) = \sum_{0}^{\infty} r(kT)e^{-skT}.$$

As sampling time, $T$, is constant, we may change the variable of interest to: $z = e^{sT}$, where $z$ similarly represents a complex variable, to represent the sampled signal as:

$$r(z) = \sum_{0}^{\infty} r(kT)z^{-k}.$$

In the above expression, where $r(z)$ defines the $z$-transform of a sampled signal, $r(kT)$, that is represented by a sequence of numerical values.

### 7.1.1 Z-transform

The $z$-transform (see Appendix) is the sampled-data equivalent of the Laplace transform used with continuous-time signals. The $z$-transform is related to the discrete-time Fourier transform (DTFT) just as the Laplace transform is related to the Fourier transform (FT). The domain of the $z$-transform includes real- or complex-valued number sequences.

To proceed further, we assume that a number sequence, $r(kT)$, is obtained by sampling a real-valued signal $r(t)$. The time dependence may be suppressed to represent the sampled sequence as: $r(k) = \{r(0), r(1), \ldots\}$. Then, the transformed sequence is given as:

$$r(z) = z[r(kT)] = \sum_{0}^{\infty} r(k)z^{-k}.$$

We note that the transformed sequence, $r(z)$, represents a rational function of a complex variable $(z)$, where $z = |z|e^{j\theta}$.

As an example, the $z$-transform of a unit-step sequence, given as: $u(k) = \{1, 1, \ldots\}$, is computed as:

$$u(z) = \sum_0^\infty z^{-k} = \frac{1}{1 - z^{-1}}; \quad |z^{-1}| < 1.$$

The convergence of the above geometric series is conditioned on satisfying: $|z^{-1}| < 1$, which defines its region of convergence (ROC). The ROC for $z^{-1}$ in the above series corresponds to the inside of a unit circle, defined by: $e^{j\theta}$, $0 \le \theta < 2\pi$, centered at the origin in the complex $z$-plane.

Further examples of $z$-transform of sampled signals are given below:

**Example 7.1:** Let $r(t) = e^{-at}u(t)$; then

$$r(kT) = \{1, \ e^{-aT}, e^{-2aT}, \ldots\}$$

$$r(z) = \sum_0^\infty e^{-akT}z^{-k} = \frac{1}{1 - e^{-aT}z^{-1}}; \ |e^{-aT}z^{-1}| < 1$$

**Example 7.2:** Let $r(t) = e^{j\omega t}u(t)$; then

$$r(kT) = \{1, e^{j\omega T}, e^{2j\omega T}, \ldots\}$$

$$r(z) = \sum_0^\infty e^{jk\omega T}z^{-k} = \frac{1}{1 - e^{j\omega T}z^{-1}}; \ |e^{j\omega T}z^{-1}| = |z^{-1}| < 1.$$

The $z$-transforms of the sampled sinusoidal signals are obtained by applying the Euler's identity to $z$-transform of $e^{jk\omega T}$ and separating the real and imaginary parts. The results are:

$$\sin(k\omega T) \overset{z}{\leftrightarrow} \frac{\sin(\omega T)z}{z^2 - 2\cos(\omega T) + 1}$$

$$\cos(k\omega T) \overset{z}{\leftrightarrow} \frac{z(z - \cos(\omega T))}{z^2 - 2\cos(\omega T) + 1}$$

$$e^{-akT}\sin(k\omega T) \overset{z}{\leftrightarrow} \frac{e^{-aT}\sin(\omega T)z}{z^2 - 2\cos(\omega T)e^{-aT} + e^{-2aT}}$$

$$e^{-akT}\cos(k\omega T) \overset{z}{\leftrightarrow} \frac{z(z - e^{-aT}\cos(\omega T))}{z^2 - 2\cos(\omega T)e^{-aT} + e^{-2aT}}.$$

We note from the above examples that in each case, the $z$-transform representation includes the specified ROC. In fact, $z$-transform description is incomplete without specifying the associated ROC.

We also note that although the definition of $z$-transform involves the Laplace transform variable (through $z = e^{sT}$), its actual computation has no reference to $s$.

The $z$-transforms of other complex signals may be obtained by using its properties of linearity, differentiation, and translation, etc. (see Appendix).

## 7.1.2 Zero-Order Hold

The zero-order hold (ZOH) represents a model of the digital-to-analog converter (DAC) that reconstructs a continuous-time signal from its sampled values. Specifically, it is a device that holds its output constant for one sampling period, $T$.

Assuming that the input to ZOH is a sampled signal, $r(kT) = r(t)|_{t=kT}$, its output is defined as:

$$r(t) = r((k-1)T) \quad \text{for} \quad (k-1)T \le t < kT.$$

The output of the ZOH to an arbitrary sampled input, $r(kT)$, is a staircase version of the analog signal, $r(t)$, delayed in time by an amount, $T/2$ (as can be confirmed by visualizing the input as a ramp signal).

The impulse response of the ZOH is given as: $G_{ZOH}(t) = 1$, $0 < t < 1$ (Figure 7.2).

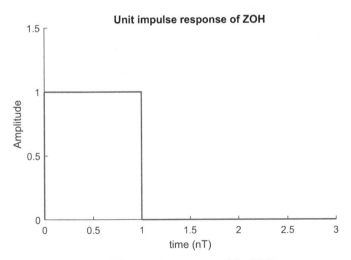

**Figure 7.2** Impulse response of the ZOH.

By applying the Laplace transform, its impulse transfer function is obtained as:

$$G_{ZOH}(s) = \frac{1}{s} - \frac{e^{-sT}}{s} = \frac{1 - e^{-sT}}{s}$$

Further, the frequency response of the ZOH is computed as:

$$G_{ZOH}(\omega) = \frac{1 - e^{-j\omega T}}{j\omega} = T\frac{\sin(\omega T/2)}{\omega T/2}e^{-j\omega T/2}$$

We note that the frequency response of the ZOH contains a *sinc* function with a delay term that contributes additional phase to the loop transfer function when the ZOH is placed in the feedback loop. The added phase from the ZOH reduces the available phase margin (Section 5.1.3); hence, the presence of ZOH adversely impacts the stability of the closed-loop system.

We may recall that the phase margin (PM) can be read from the Bode phase plot of the loop transfer function: $KGH(j\omega)$. In particular, let $\omega_{gc}$ denote the gain crossover frequency, that is, the frequency at which the Bode magnitude plot has unity gain: $|KGH(j\omega_{gc})|_{dB} = 0dB$; then, the PM is computed as: $PM = 180° - \angle KGH(j\omega_{gc})$.

The presence of ZOH in the feedback loop reduces the phase margin (PM) by an amount $\omega_{gc}T/2$. Hence, the desired sampling time $T$ can be obtained from the acceptable degradation in the PM accompanying the ZOH. For example, if we want to limit the PM degradation to 5° ($0.087 \ rad$), the sampling period should be selected in accordance with: $T \leq \frac{0.175}{\omega_{gc}}$.

### 7.1.3 Pulse Transfer Function

To analyze a sampled-data system that includes a continuous-time plant driven by a digital controller through the ZOH, we apply the $z$-transform to obtain a transfer function description of the plant cascaded with the ZOH. The resulting pulse transfer function, $G(z)$, describes the input–output behavior of the plant at the sampling intervals.

The pulse transfer function of a continuous-time plant, $G(s)$, is obtained as:

$$G(z) = z\left\{\frac{1 - e^{-sT}}{s}G(s)\right\} = (1 - z^{-1})z\left\{\frac{G(s)}{s}\right\}.$$

where $z\{\cdot\}$ denotes the $z$-transform of a sequence obtained by sampling a physical signal that represents the time integral of the impulse response: $\int g(t)dt$.

The conversion from Laplace domain to $z$-domain involves intermediate steps, that is, inverse Laplace transform followed by signal sampling to obtain a number sequence, and the application of $z$-transform. The $z$-transform tables obviate this difficulty by providing $z$-transform and Laplace transform of representative time signals side by side (see Appendix).

The computation of the pulse transfer function for given plant models is illustrated through the following examples.

**Example 7.3:** A first-order system.

Let $G(s) = \frac{a}{s+a}$; then

$$G(z) = (1 - z^{-1})\, z \left\{ \frac{a}{s(s+a)} \right\} = (1 - z^{-1}) z \left\{ \frac{1}{s} - \frac{1}{s+a} \right\}.$$

From the $z$-transform table (see Appendix),

$$G(z) = \frac{z-1}{z} \left( \frac{z}{z-1} - \frac{z}{z - e^{-aT}} \right) = \frac{1 - e^{-aT}}{z - e^{-aT}}.$$

**Example 7.4:** A second-order system with a pole at the origin.

Let $G(s) = \frac{a}{s(s+a)}$; then

$$G(z) = (1 - z^{-1})\, z \left\{ \frac{a}{s^2(s+a)} \right\} = (1 - z^{-1}) z \left\{ \frac{1}{s^2} - \frac{1/a}{s} + \frac{1/a}{s+a} \right\}.$$

Using the $z$-transform table,

$$G(z) = \frac{z-1}{z} \left[ \frac{Tz}{(z-1)^2} - \frac{1}{a}\left( \frac{z}{z-1} - \frac{z}{z - e^{-aT}} \right) \right]$$

$$= \frac{aT(z - e^{-aT}) - (z-1)(1 - e^{-aT})}{a(z-1)(z - e^{-aT})}.$$

From the above examples, we make the following observations:

1. The order of the pulse transfer function, that is, the degree of the denominator polynomial in $G(z)$, matches that of the continuous-time transfer function, $G(s)$.
2. The poles of the pulse transfer function are related to the continuous-time plant poles through $z_i = e^{s_i T}$.
3. The pulse transfer functions of the second- and higher-order systems additionally include finite zeros.

**Pulse Transfer Function in MATLAB.** The pulse transfer function, $G(z)$, for a given plant transfer function, $G(s)$, can be easily obtained in the MATLAB Control Systems Toolbox by using the "*c2d*" command and specifying a sampling time $(T_s)$. In addition to the ZOH (default), the *c2d* command allows a choice of an input method from the following choices: "zoh", "foh", "impulse", "matched", "tustin", and 'least-squares'.

The input argument to the *c2d* command is a dynamic system object (DSO) created in the Control Systems Toolbox using one of the following commands: "tf", "ss", "zpc", and "frd".

Given a continuous-time transfer function, the MATLAB commands to obtain the pulse transfer function are invoked as follows:

```
G=tf(num, den);          % define plant transfer function
Gz=c2d(G, T)             % obtain pulse transfer function
```

The choice of the sampling time is guided by the natural frequency of the system, or equivalently, the magnitude of the dominant poles of the system. In particular, we may choose $T_s \ll 1/\omega_n$, where $\omega_n$ represents the natural frequency. This is illustrated in the following examples.

**Example 7.5:** Let $G(s) = \frac{1}{s+1}$; $T = 0.2s$. Then, the pulse transfer function is obtained from MATLAB as: $G(z) = \frac{0.181}{z-0.819}$. We may note that, $G(z) = \frac{1-e^{-0.2}}{z-e^{-0.2}}$.

**Example 7.6:** Let $G(s) = \frac{1}{s(s+1)}$, $T = 0.2s$; then, the pulse transfer function is obtained from MATLAB as: $G(z) = \frac{0.0187(z+0.936)}{(z-1)(z-0.819)}$.

**Pulse Transfer Function in the MATLAB Symbolic Math Toolbox.** The pulse transfer function, $G(z)$, of a continuous-time plant described by, $G(s)$, can be analytically obtained in the MATLAB Symbolic Math Toolbox by first obtaining its unit-impulse response, followed by sampling at $T$sec intervals and the application of $z$-transform.

As an example, the pulse transfer function for a first-order plant (Example 7.5) is obtained by issuing the following MATLAB commands:

```
syms s z                          % define complex variables
syms t a T real                   % define real-valued variables
syms n integer                    % define integer-valued variables
sympref('FloatingPointOutput',true); % define output format
G=1/(s+1);                        % define plant transfer function
g=ilaplace(G/s);                  % inverse Laplace transform
gn=subs(g,t,n*T);                 % discretize the time function
Gz=(z-1)/z*ztrans(gn);            % obtain pulse transfer function
Gz=subs(Gz,'T',0.2);              % specify sampling time
Gz=simplify(Gz)                   % simplify the result
```

Similarly, the pulse transfer function of a second-order system (Example 7.6) is obtained by issuing the following MATLAB commands after declaring the symbolic variables:

```
G=1/s/(s+1);              % define plant transfer function
g=ilaplace(G/s);          % inverse Laplace transform
gn=subs(g,t,n*T);         % discretize the time function
Gz=(z-1)/z*ztrans(gn);    % obtain pulse transfer function
Gz=subs(Gz,'T',0.2);      % specify sampling time
Gz=simplify(Gz)           % simplify the result
```

## 7.2 Sampled-Data System Response

The response of a sampled-data system, described by its pulse transfer function, $G(z)$, to an input sequence, $u(k)$, is represented as the output sequence, $y(k)$. In the $z$-domain, the system response is obtained as: $y(z) = G(z)u(z)$.

Alternatively, we may use inverse $z$-transform to obtain an input–output description in the form of a difference equation that can be iteratively solved for a given input sequence to obtain the output sequence. This approach is described next.

### 7.2.1 Difference Equation Solution by Iteration

A difference equation description of the plant at sampling intervals is obtained from its pulse transfer function. Accordingly, let $\frac{y(z)}{u(z)} = G(z) = \frac{n(z)}{d(z)}$, where $n(z)$ and $d(z)$ are, respectively, $m$th order and $n$th order polynomials, given as:

$$n(z) = \sum_{i=0}^{m} b_i z^{m-i},$$

$$d(z) = z^n + \sum_{i=1}^{n} a_i z^{n-i}.$$

First, we cross-multiply $\frac{y(z)}{u(z)} = \frac{n(z)}{d(z)}$ to get: $d(z)y(z) = n(z)u(z)$ and rearrange the result to obtain an expression for $z^n y(z)$. Next, we apply the inverse $z$-transform (see Appendix) to obtain a corresponding difference equation, given as:

$$y(k + n) = -\sum_{i=1}^{n} a_i y(k + n - i) + \sum_{i=0}^{m} b_i u(k + m - i)$$

As the system is linear and time-invariant, we can a time shift to obtain a time-domain relation of the form:

$$y(k) = -\sum_{i=1}^{n} a_i y(k - i) + \sum_{i=0}^{m} b_i u(k - i)$$

The above expression represents an update rule that can be programmed on a computer to simulate sampled-data system response to an input sequence, $u(k)$, which is obtained by sampling an analog signal $u(t)$, that is, $u(k) = u(t)|_{t=kT}$.

The iterative approach to solving for the sampled-data system output is illustrated using the following examples.

**Example 7.7:** A first-order sampled-data system (continued from Example 7.5).

Let $G(z) = \frac{y(z)}{u(z)} = \frac{0.181}{z-0.819}$; after cross-multiplying, using inverse $z$-transform, and applying a time shift, the sampled-data system is described by the following time-domain input–output relation:

$$y(k) = 0.819y(k - 1) + 0.181u(k - 1).$$

**Example 7.8:** A second-order sampled-data system (continued from Example 7.6).

Let $G(z) = \frac{0.0187(z+0.936)}{(z-1)(z-0.819)}$; then, after cross-multiplying, using inverse $z$-transform, and applying a time shift, the sampled-data system is described by the following time-domain input–output relation:

$$y(k) = 1.819y(k - 1) - 0.819y(k - 2) + 0.0187u(k - 1) + 0.0175u(k - 2).$$

In particular, we are interested in the unit-pulse and unit-step responses of the sampled-data systems, which are described next.

### 7.2.2 Unit-Pulse Response

The unit-pulse response of a sampled-data system described by its pulse transfer function $G(z)$ is its response to a unit-pulse $r(k) = \delta(k)$. Since $\delta(z) = 1$, the unit-pulse response is computed from inverse $z$-transform of $G(z)$, that is, $g(kT) = z^{-1}[G(z)]$, where a sampling time of $T = 1s$ is assumed. Unit-pulse response for alternate sampling times can be obtained by assuming $\delta(z) = \frac{1}{T}$, so that $g(kT) = \frac{1}{T}z^{-1}[G(z)]$.

Alternatively, given the time-domain input–output description of a sampled-data system, its unit-pulse response is easily computed by iteration.

The unit-pulse response comprises the natural response modes of a sampled-data system. Similar to the continuous-time system models, these modes are derived from the roots of the denominator polynomial in the pulse transfer function, $G(z) = n(z)/d(z)$, and can be characterized as follows:

**Real and Distinct Roots.** We assume that the denominator polynomial, $d(z)$, has real and distinct roots: $z_i$, $i = 1, \ldots, n$; then, its natural response modes are given as: $\phi_i(k) = (z_i)^k, i = 1, \ldots, n$. We may note that these terms die out with time if $|z_i| < 1$, that is, if the root lies inside the unit-circle.

**Complex Roots.** We assume that the denominator polynomial, $d(z)$, has complex roots of the form: $re^{\pm j\theta}$; then, its natural response modes are: $\phi_i(k) = \{r^k \cos k\theta, \ r^k \sin k\theta\}$. We note that these modes similarly die out with time if $|r| < 1$, that is, if the complex roots are inside the unit circle.

The commutation of unit-pulse response is illustrated through the following examples:

**Example 7.9:** A first-order sampled-data system (continued from Example 7.7).

Let $G(z) = \frac{y(z)}{u(z)} = \frac{0.181}{z-0.819}$, where a sampling time: $T = 0.2s$ was used. The sampled-data system is described by the following time-domain relation:

$$y(k) = 0.819y(k-1) + 0.181u(k-1).$$

Since a sample time $T = 0.2s$ was assumed, the discretized impulse input is represented as: $\delta(k) = \{5, 0, 0, \ldots\}$. Assuming zero initial conditions, the output is obtained by iteration. The following MATLAB script can be used to obtain the first few terms of the unit-pulse response sequence:

```
y=zeros(1,10);                    % intialize response array
for n=2:10, y(n)=.819*y(n-1)+0.906*(n==2); end % iterative loop
```

The resulting impulse response sequence is given as:

$$y(k) = \{0, \ 0.906, \ 0.742, \ 0.608, \ 0.497, 0.407, \ 0.333, \ \ldots\}$$

**Example 7.10:** A second-order sampled-data system (continued from Example 7.8).

Let $G(z) = \frac{0.0187(z+0.936)}{(z-1)(z-0.819)}$, $T = 0.2s$. Then, the sampled-data system is described by the following time-domain input–output relation:

$$y(k) = 1.819y(k-1) - 0.819y(k-2) + 0.0187u(k-1) + 0.0175u(k-2).$$

Next, using $u(k) = \{5, 0, 0, \ldots\}$, and assuming zero initial conditions, the output is obtained by iteration. The following MATLAB commands are used to obtain the first few terms of the impulse response:

```
y=zeros(1,10);               % intialize response array
y(2)=.01873*5;               % specify initial condition
y(3)=1.819*y(2)+.0175*5;     % compute initial value
for n=4:10, y(n)=1.819*y(n-1)-.819*y(n-2); end % iterative loop
```

The resulting impulse response sequence is given as:

$$y(k) = \{0,\ 0.0937,\ 0.258,\ 0.392,\ 0.502, 0.593,\ 0.667,\ \ldots\}$$

**Unit-Pulse Response in MATLAB.** The "impulse" command in the MAT-LAB Control Systems Toolbox can be used to obtain and/or plot the unit-pulse response of a sampled-data system.

The MATLAB commands to obtain the unit-pulse response of a transfer function model are given below:

```
G=tf(num, den);      % define plant transfer function
Gz=c2d(G, T);        % obtain pulse transfer function
y=impulse(Gz);       % obtain impulse response sequence
impulse(Gz)          % plot the impulse response
```

For the first-order system, let $G(z) = \frac{0.181}{z-0.819}$; then, by using the above MATLAB commands, the unit-pulse response sequence is obtained as:

$$g(k) = \{0,\ 0.906,\ 0.742,\ 0.608,\ 0.497, 0.407,\ 0.333,$$
$$0.273,\ 0.224\ldots\}$$

For the second-order system, let $G(z) = \frac{0.0187(z+0.936)}{(z-1)(z-0.819)}$; then, from MATLAB the unit-pulse response sequence is obtained as:

$$g(k) = \{0,\ 0.094,\ 0.258,\ 0.392,\ 0.503,\ 0.593,\ 0.667,$$
$$0.727,\ 0.777, \ldots\}.$$

The unit-pulse response sequences for the first- and second-order systems obtained from MATLAB match those obtained previously through iteration. The unit-pulse responses are plotted in Figure 7.3. The impulse response of the continuous-time system is plotted alongside for comparison.

We note that for the second-order system, the integrator in the loop causes the impulse response for both the continuous-time and discrete-time systems to asymptotically approach unity in the steady-state.

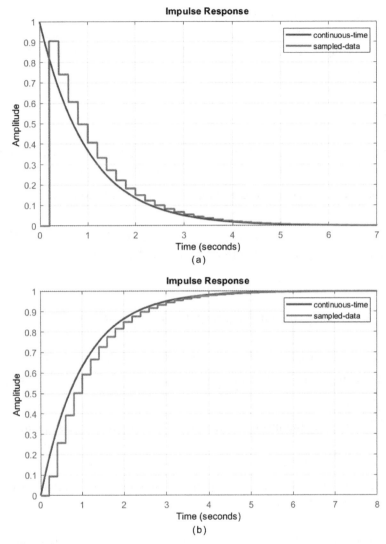

**Figure 7.3** The unit-pulse response of a sampled-data system: First-order system (a); Second-order system (b).

### 7.2.3 Unit-Step Response

The unit-step response of the sampled-data system is its response to the unit-step sequence: $r\{kT\} = \{1, 1, 1 \ldots\}$; $r(z) = \frac{1}{1-z^{-1}}$. The unit-step response can be computed in the time-domain by iteration or in the $z$-domain by solving for the output as: $y(z) = \frac{G(z)}{1-z^{-1}}$. The inverse $z$-transform is then applied to obtain $y(k)$.

**Unit-Step Response in MATLAB.** The MATLAB "step" command from the Control Systems Toolbox can be used to obtain and/or plot the step response of the sampled-data system. The relevant MATLAB commands are given below:

```
G=tf(num, den);            % define plant transfer function
Gz=c2d(G, T);              % obtain pulse transfer function
y=step(Gz);                % obtain step response sequence
step(Gz)                   % plot the step response
```

The computation and plotting of the unit-step response in the case of sampled-data systems is illustrated through the following examples:

**Example 7.11:** A first-order sampled-data system (continued from Example 7.9).

Let $G(s) = \frac{1}{s+1}$; we use $T = 0.2s$ to obtain: $G(z) = \frac{0.181}{z-0.819}$. The sampled-data system is described by a time-domain input–output relation:

$$y(k) = 0.819y(k-1) + 0.181u(k-1).$$

Analytically, the unit-step response is computed as: $y(z) = \frac{G(z)}{1-z^{-1}}$, or $y(z) = \frac{0.181z}{(z-1)(z-0.819)}$.

In order to take the inverse $z$-transform of the output we first expand $y(z)/z$ in partial fractions and multiply by $z$ to obtain: $y(z) = \left( \frac{z}{z-1} - \frac{z}{z-0.819} \right)$.

Then, using the inverse $z$-transform, the step response sequence is given as:

$$y(kT) = 1 - (0.819)^k, \quad k = 0, 1, \ldots$$

For comparison, the step response of an equivalent continuous-time system is obtained as follows: let $G(s) = \frac{1}{s+1}$; then, for $r(t) = u(t)$; we have: $y(t) = 1 - e^{-t}, \quad t > 0$.

The step responses of the continuous and discrete systems are plotted alongside (Figure 7.4(a)). The MATLAB commands for this example are given as follows:

```
G=tf(1,[1 1]);             % define plant transfer function
Gz=c2d(G, 0.2);            % obtain pulse transfer function
step(G,Gz),grid            % plot step response
legend('continuous-time','sampled-data') % figure legend
```

We note that the step response of first-order continuous-time and discrete-time systems both asymptotically reach a value of unity.

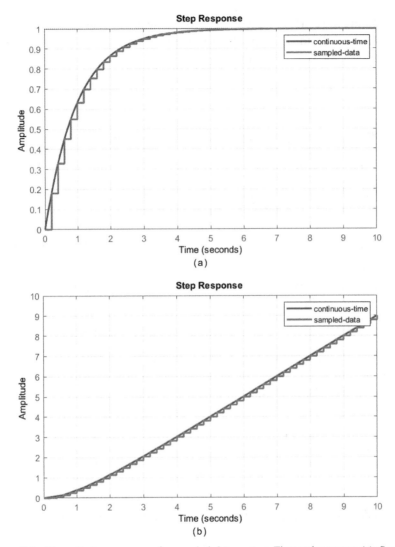

**Figure 7.4** The unit-step response of a sampled-data system: First-order system (a); Second-order system (b).

**Example 7.12:** A second-order sampled-data system (continued from Example 7.10).

Let $G(s) = \frac{1}{s(s+1)}$; we use $T = 0.2s$ to obtain:

$$G(z) = \frac{0.0187z + 0.0175}{z^2 - 1.819z + 0.819}.$$

Using the time-domain relation for the second-order sampled-data system (Example 7.8), the first few terms of the step response are computed by using the following MATLAB commands:

```
y=zeros(1,10);                    % intialize step response array
y(2)=0.0187;                      % set intial value
y(3)=2.819*0.0187+.0175;          % set intial value
for k=4:10, y(k)=1.819*y(k-1)-0.819*y(k-2)+0.0363; end % loop
```

The resulting unit-step response sequence is given as:

$$\{0, \quad 0.0187, \quad 0.0702, \quad 0.149, \quad 0.249, \quad 0.368, \quad 0.501, 0.647, 0.802, 0.965, \ldots\}$$

The above unit-step response continues to increase out of bound due to the presence of an integrator in the plant model.

Analytically, the unit-step response of the sampled-data system may be obtained as follows: let $r(z) = \frac{z}{z-1}$ (unit-step); then, the output is given as:

$$y(z) = \frac{z(0.0187z+0.0175)}{(z-1)(z^2-1.819z+0.819)}.$$

In order to apply the inverse $z$-transform, we use the long division to express the quotient as:

$$y(z) = 0.0187z^{-1} + 0.0703z^{-2} + 0.149z^{-3} + 0.249z^{-4} + 0.501z^{-5} + \cdots$$

The resulting output sequence after application of the inverse $z$-transform matches the one obtained by iteration. The following MATLAB commands were used to obtain the quotient:

```
G=tf(1,[1 1 0]);                  % define plant transfer function
Gz=c2d(G, .2);                    % pulse transfer function (T=.2s)
[num,den]=tfdata(Gz);             % obtain numerator and denominator
num=[num{:} zeros(1,10)];         % add zeros to numerator polynomial
den=conv(den{:},[1 -1]);          % add integrator to denominator
q=deconv(num,den)                 % compute the quotient
```

For comparison with an equivalent continuous-time system, we may consider an analog plant: $G(s) = \frac{1}{s(s+1)}$. Then, for input, $r(s) = \frac{1}{s}$, we have: $y(s) = \frac{1}{s^2(s+1)} = \frac{1}{s^2} - \frac{1}{s} + \frac{1}{s+1}$, or, in the time-domain,

$$y(t) = (t - 1 + e^{-t})u(t).$$

The step responses of the continuous-time and discrete-time systems are plotted alongside (Figure 7.4(b)). The MATLAB commands for this example are given as follows:

```
G=tf(1,[1 1 0]);                  % define plant transfer function
Gz=c2d(G, .2);                    % pulse transfer function (T=.2s)
step(G,Gz),grid                   % plot step response
legend('continuous-time','sampled-data') % figure legend
```

### 7.2.4 Response to Arbitrary Inputs

Given the unit-pulse response of the system, its response to an arbitrary input sequence is obtained by convolution. Accordingly, let the unit-pulse response sequence be given as:

$$\{g(k)\} = \{g(1),\ g(2),\ \ldots\}$$

Then, using an arbitrary input sequence: $\{u(k)\} = \{u(1),\ u(2),\ \ldots\}$, the output sequence is given by the convolution sum:

$$y(k) = g(k) * u(k) = \sum_{i=0}^{\infty} g(k-i)u(i)$$

As an example, we obtain the output for a sinusoidal input below.

**Example 7.13:** A second-order sampled-data system.

Let $G(z) = \frac{0.368z+0.264}{(z-1)(z-0.368)}$; we assume a sinusoidal input, $u(kT) = \sin(\pi kT)$.

Then, the output response is computed using the following MATLAB commands (Figure 7.5):

```
G=tf(1,[1 1 0]);              % define plant transfer function
Gz=c2d(G,.2);                 % obtain pulse transfer function
g=impulse(ss(Gz));            % compute impulse response
t=0:.2:5;                     % define time range
u=sin(pi*t);                  % define input
y=conv(g,u);                  % perform convolution
y1=lsim(G,u,t);               % simulate contnuous-time system
plot(t,5*y1), grid,hold       % plot cont-time system response
stairs(t,y(1:26)), hold       % plot discrete system response
legend('continuous-time','sampled-data')  % figure legend
```

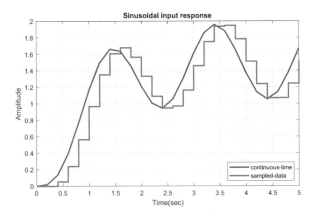

**Figure 7.5**  Sinusoidal response of a second-order sampled-data system preceded by a ZOH.

## 7.3 Stability in the Case of Sampled-Data Systems

We recall that stability in the case of continuous-time systems requires that the closed-loop system poles (roots of the characteristic polynomial, $\Delta(s)$) are confined to the OLHP, where $s = j\omega$ defines the stability boundary.

The $z$-plane stability boundary is obtained from the transform: $z = e^{j\omega T} = 1\angle\omega T$, which maps the $j\omega$-axis to the unit circle in the complex $z$-plane, and the OLHP to the inside of the unit circle: $|z| < 1$.

Accordingly, the closed-loop sampled-data system is stable if and only if the pulse characteristic polynomial $\Delta(z)$ has its roots inside the unit circle, that is, if $z_i$ is the root of $\Delta(z)$, then the stability requirement is: $|z_i| < 1$.

The analytical conditions for a $z$-domain polynomial, $A(z)$, to have its roots inside the unit circle are given by the Schur–Cohn stability test (see References). When adapted to real polynomials, the Schur–Cohn test results in a criterion similar to the Ruth's test and is known as Jury's stability test.

### 7.3.1 Jury's Stability Test

To perform the Jury's stability test, we assume that the $n$th order $z$-domain polynomial to be investigated is given as:

$$A(z) = a_0 z^n + a_1 z^{n-1} + \cdots + a_{n-1} z + a_n; a_0 > 0.$$

Next, we build the Jury's table, which is given as:

$$\begin{vmatrix} a_n & a_{n-1} & \cdots & a_0 \\ a_0 & a_1 & \cdots & a_n \\ b_{n-1} & b_{n-2} & \cdots & b_0 \\ b_0 & b_1 & \cdots & b_{n-1} \\ c_{n-2} & c_{n-1} & \cdots & c_0 \\ c_0 & c_1 & \cdots & c_{n-2} \\ & & \vdots & \end{vmatrix}$$

where the coefficients of the third and subsequent rows are computed as:

$$b_k = \begin{vmatrix} a_n & a_{n-1-k} \\ a_0 & a_{k+1} \end{vmatrix}, \quad k = 0, \ldots, n-1$$

$$c_k = \begin{vmatrix} b_{n-1} & b_{n-2-k} \\ b_0 & b_{k+1} \end{vmatrix}, \quad k = 0, \ldots, n-2, \text{ etc.}$$

The necessary conditions for polynomial stability are:

$$A(1) > 0, (-1)^n A(-1) > 0.$$

The sufficient conditions for stability, given by the Jury's test, are:

$$a_0 > |a_n|, \ |b_{n-1}| > |b_0|, \ |c_{n-2}| > |c_0|, \ldots \ (n-1 \text{ conditions})$$

**Stability Test for a Second-Order Polynomial.** In the case of a second-order polynomial, $A(z) = z^2 + a_1 z + a_2$, the Jury's table is given as:

$$\begin{vmatrix} a_2 & a_1 & 1 \\ 1 & a_1 & a_2 \\ b_1 & b_0 & \\ b_0 & b_1 & \end{vmatrix}$$

The resulting necessary conditions are: $1 + a_1 + a_2 > 0, 1 - a_1 + a_2 > 0$. The sufficient conditions are: $|a_2| < 1, |1 + a_2| > |a_1|$.

## 7.3.2 Stability Through Bilinear Transform

The bilinear transform (BLT) defines a linear map from the $s$-domain to the $z$-domain and vice versa. In order to develop the BLT, we use the first-order Pade' approximation for $z = e^{sT}$ to obtain:

$$z = \frac{e^{sT/2}}{e^{-sT/2}} \cong \frac{1 + sT/2}{1 - sT/2}, \quad s = \frac{2}{T} \frac{z-1}{z+1}$$

Since $T$ has no impact on stability determination, we may use $T = 2$ for simplicity. Further, in order to distinguish the characteristic polynomial obtained through BLT from the original $\Delta(s)$, a new complex variable, $w$, is introduced. The updated BLT is given as:

$$z = \frac{1 + w}{1 - w}, \quad w = \frac{z-1}{z+1}$$

Application of the above BLT to $\Delta(z)$ returns the characteristic polynomial $\Delta(w)$, whose stability is then determined through the application of Hurwitz criterion (Section 2.5).

**Stability of a Second-Order Polynomial.** In particular, we consider a second-order polynomial: $\Delta(z) = z^2 + a_1 z + a_2$. The application of BLT, ignoring the denominator term in the result, gives:

$$\Delta(w) = \Delta(z)|_{z=\frac{1+w}{1-w}} = (1 - a_1 + a_2)w^2 + 2(1 - a_2)w + 1 + a_1 + a_2$$

Then, application of the Hurwitz criterion for second-order polynomial results in the following stability conditions for $\Delta(z)$:

$$a_2 + a_1 + 1 > 0, \quad a_2 - a_1 + 1 > 0, \quad 1 - a_2 > 0$$

These conditions match those obtained by the application of Jury's stability test.

Examples of stability determination are given in Examples 7.14–7.16 below.

## 7.4 Closed-Loop Sampled-Data Systems

A closed-loop sampled-data control system contains a comparator, a sampler, a digital controller, a ZOH, a continuous-time plant, and a measurement sensor (Figure 7.1). In particular, we consider a unity gain feedback control system, where a continuous-time plant is driven by a digital controller through a ZOH. Further, we consider a static controller represented by a scalar gain, $K$. Then, the closed-loop pulse transfer function is given as:

$$T(z) = \frac{y(z)}{r(z)} = \frac{KG(z)}{1 + KG(z)}.$$

Let $G(z) = \frac{n(z)}{d(z)}$; then we have: $T(z) = \frac{Kn(z)}{d(z) + Kn(z)}$.

### 7.4.1 Closed-Loop System Stability

The closed-loop system stability can be determined by applying the Jury's stability test to the closed-loop characteristic polynomial: $\Delta(z) = 1 + KG(z)$. Alternatively, we may use the BLT to determine stability. The stability determination is illustrated using the following examples.

**Example 7.14:** Let $\Delta(z) = z + 0.632K - 0.368$; then $\Delta(z)$ is stable for $|0.632K - 0.368| < 1$, or $-1 < K < 2.16$.

**Example 7.15:** Let $\Delta(z) = z^2 + z + K$; then, from Jury's stability test we obtain the following stability condition: $|K| < 1$. Alternately, using BLT, we obtain the characteristic polynomial as:

$$\Delta(w) = \Delta(z)|_{z = \frac{1+w}{1-w}} = Kw^2 + 2(1 - K)w + 1 + K$$

The application of the Hurwitz criteria to $\Delta(w)$ reveals $0 < K < 1$ for stability.

**Example 7.16:** Let $G(s) = \frac{1}{s(s+1)}$, $T = 0.2s$; then, the pulse transfer function is: $G(z) = \frac{0.0187z + 0.0175}{(z-1)(z-0.181)}$. Further, the characteristic polynomial is: $\Delta(z) = z^2 + (0.0187K - 1.819)z + 0.0175K + 0.819$.

The $w$-polynomial obtained through BLT is given as:

$$\Delta(w) = (3.637 - 0.0012K)w^2 + (0.363 - 0.035K)w + 0.036K$$

The application of the Hurwitz criteria to $\Delta(w)$ gives $0 < K < 10.34$ for stability.

The above results can be obtained by using the following commands from MATLAB Control Systems and Symbolic Math Toolboxes.

```
syms z w                    % declare symbolic variables
syms K real                 % declare symbolic gain
G=tf(1,[1 1 0]);            % define dynamic system
Gz=c2d(G,.2);               % discretize system
[num,den]=tfdata(Gz);       % obtain numerator and denominator
pz=poly2sym(den{:},z);      % use symbolic variables
pz=pz+K*poly2sym(num{:},z); % characteristic polynomial
pw=subs(pz,z,(1+w)/(1-w));  % apply BLT
pw=simplify(pw*(w-1)^2);    % simplify result
cw=coeffs(pw,w);            % obtain coefficients
solve(cw(2),K)              % solve for K
```

### 7.4.2 Unit-Step Response

Given the closed-loop pulse transfer function, $T(z)$, we can compute its response to a unit-step sequence, $r(kT) = \{1, 1, \ldots\}$. The unit-step response can be obtained by iteration, or analytically from the expression: $y(z) = T(z)r(z)$.

The unit-step response is explored in the case of first- and second-order system below.

**Example 7.17:** A first-order sampled-data system (continued from Example 7.11).

Let $G(s) = \frac{1}{s+1}$; and, let $T = 0.2s$; then, $G(z) = \frac{0.181}{z-0.819}$. Next, for a static gain controller, $K$, the closed-loop pulse transfer function is given as: $T(z) = \frac{0.6181K}{z-0.819+0.181K}$.

For simplicity, we assume $K = 1$; further, let $r(kT) = 1, r(z) = \frac{1}{1-z^{-1}}$; then, in the $z$-domain the system output is given as: $y(z) = \frac{0.181z}{(z-1)(z-0.638)}$.

In order to take the inverse $z$-transform, we expand $y(z)/z$ in partial fractions, then multiply the terms by $z$ to obtain:

$$y(z) = 0.5 \left( \frac{z}{z-1} - \frac{z}{z-0.638} \right).$$

Taking inverse $z$-transform, the sampled output sequence is given as:

$$y(kT) = 0.5(1 - 0.638^k), \quad k = 0, 1, \ldots$$

The resulting output sequence is:

$$\{0, \ 0.632, \ 0.465, \ 0.509, \ 0.498, \ 0.501, \ 0.5, \ 0.5 \ldots\}.$$

In order to compare the unit-step response of the sampled-data system with that of a continuous-time system, let: $G(s) = \frac{1}{s+1}, H(s) = 1$; then, we have: $T(s) = \frac{K}{s+1+K}$. Let $r(t) = u(t)$; and assume $K = 1$; then, the continuous-time system output is given as: $y(t) = \frac{1}{2}(1 - e^{-2t}), t > 0$.

The step response of the sampled-data and the continuous-time systems are plotted using the following MATLAB commands (Figure 7.6(a)):

```
G=tf(1,[1 1])                  % define plant transfer function
Gz=c2d(G, .2)                  % obtain pulse transfer function
T=feedback(G,1);               % obtain closed-loop system
Tz=feedback(Gz,1);             % obtain closed-loop system
step(T,Tz),grid                % plot step response
legend('continuous-time','sampled-data ') % figure legend
```

**Example 7.18:** A second-order sampled-data system (continued from Example 7.12).

Let $G(s) = \frac{1}{s(s+1)}$; we use $T = 0.2s$ to obtain: $G(z) = \frac{0.0187z + 0.0175}{(z-1)(z-0.819)}$.
The closed-loop pulse transfer function is given as:

$$T(z) = \frac{(0.0187z + 0.0175)K}{(z-1)(z-0.819) + (0.0187z + 0.0175)K}$$

We assume a controller gain, $K = 1$; then, the closed-loop pulse transfer function is given as: $T(z) = \frac{0.0187z + 0.0175}{z^2 - 1.8z + 0.836}$. Let $r(z) = \frac{z}{z-1}$ (unit-step); then, the system response is given as:

$$y(z) = \frac{z(0.0187z + 0.0175)}{(z-1)(z^2 - 1.8z + 0.836)}.$$

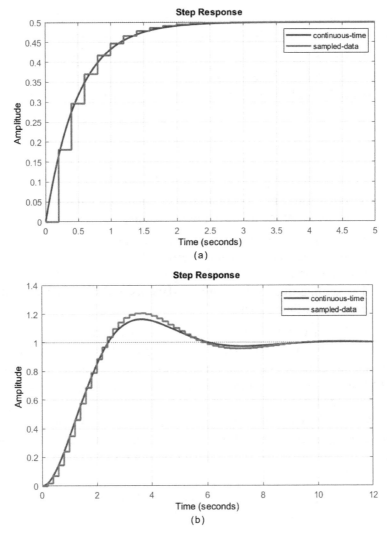

**Figure 7.6**   The unit-step response of a sampled-data system: First-order system (a); Second-order system (b).

In order to apply the inverse $z$-transform, we use the long division to express the quotient as:

$$y(z) = 0.368z^{-1} + z^{-2} + 1.4z^{-3} + 1.4z^{-4} + 1.15z^{-5} + \cdots$$

Then, the step response at sampling intervals ($T = 1$s) is given by the following sequence:

$$y(kT) = \{0, \ 0.368, \ 1, \ 1.4, \ 1.4, \ 1.15, \ \ldots\}$$

For comparison, the closed-loop transfer function for the corresponding continuous-time plant is given as: $T(s) = \frac{1}{s^2+s+1}$. Its unit-step response is obtained as: $y(s) = \frac{1}{s(s^2+s+1)} = \frac{1}{s} - \frac{1}{s^2+s+1}$, which transforms into: $y(t) = (1 - 1.15e^{-0.5t}\sin 0.866\,t)u(t)$.

The MATLAB script for this example is given as follows (Figure 7.6(b)):

```
G=tf(1,[1 1 0])                % define plant transfer function
Gz=c2d(G, 0.2)                 % pulse transfer function (T=0.2)
T=feedback(G,1);               % obtain closed-loop system
Tz=feedback(Gz,1);             % obtain closed-loop system
step(T,Tz),grid                % plot step response
legend('continuous-time','sampled-data') % figure legend
```

By comparing the continuous-time and discrete-time system step responses, we make the following observations:

1. While the continuous-time system step response with $\zeta = 0.5$ has a $16.3\%$ overshoot (Figure 7.6(b)), the corresponding sampled-data system response has somewhat higher ($18\%$) overshoot.
2. The settling time of the sampled-data system is longer than that of the continuous-time system.

The higher overshoot in the case of sampled-data system occurs due to the negative phase contributed by the ZOH that reduces the available PM by an amount: $\Delta\phi_m = \frac{\omega_{gc}T}{2}$. For this example, the continuous-time system has a PM of $52°$ at $\omega_{gc} = 0.786\frac{rad}{s}$; hence, the reduction is phase margin is: $\Delta\phi_m = 4.5°$.

A table showing the estimated overshoot for various sampling times in the case of the above second-order sampled-data system is given below. The unit-step response of the sampled-data system for two values of $T$ is plotted (Figure 7.7):

| Sample Time (T) | Reduction in PM | % Overshoot |
|:---:|:---:|:---:|
| $0.1s$ | $2.3°$ | $16.5\%$ |
| $0.2s$ | $4.5°$ | $18\%$ |
| $0.5s$ | $11.3°$ | $30\%$ |
| $1s$ | $22.5°$ | $45\%$ |

## 7.4.3 Steady-State Tracking Error

The steady-state tracking error describes the long-term behavior of the sampled-data control systems in response to a constant or linearly time-varying reference input. Ideally, we would like to have: $\lim_{k\to\infty} e(kT) = 0$.

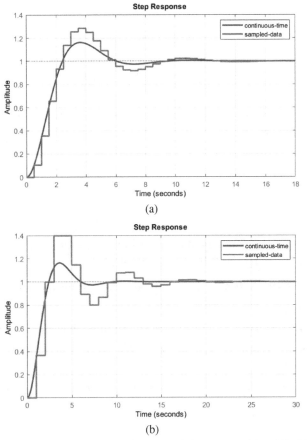

**Figure 7.7** Step response of a second-order sampled-data system: $T = 0.5s$ (a); $T = 1s$ (b).

In particular, the error, $e(z)$, in response to a given reference input, $r(z)$, in the case of a unity feedback sampled-data system is computed as:

$$e(z) = r(z)(1 - T(z)).$$

The steady-state error in the time-domain is computed by applying the final value theorem (FVT) in the $z$-domain, which is stated as:

$$\lim_{k \to \infty} e(k) = \lim_{z \to 1} (z - 1)e(z).$$

The steady-state error is commonly described using position and velocity error constants. In the case of sampled-data systems, the error constants are defined as:

$$K_p = \lim_{z \to 1} G(z)$$

$$K_v = \lim_{z \to 1} \frac{(z-1)}{T} G(z)$$

Using the error constants, the steady-state errors to step and ramp inputs are computed as:

$$e_{ss}\big|_{step} = \frac{1}{1 + K_p}$$

$$e_{ss}\big|_{ramp} = \frac{1}{K_v}.$$

Computation of steady-state error in the case of sampled-data systems is illustrated by using the following examples.

**Example 7.19:** A first-order sampled-data system (continued from Example 7.18).

Let $G(z) = \frac{0.0187z + 0.0175}{(z-1)(z-0.819)}$ $(T = 0.2s)$; then, the sample-data system error constants are given as: $K_p = \infty$, $K_v = 1$. Accordingly, $e_{ss}\big|_{step} = 0$; $e_{ss}\big|_{ramp} = 0.2$.

Alternatively, let $G(z) = \frac{0.368z + 0.264}{(z-1)(z-0.368)}$ $(T = 1s)$; then, we have $K_p = \infty$, $K_v = 1$. Accordingly, $e_{ss}\big|_{step} = 0$; $e_{ss}\big|_{ramp} = 1$.

## 7.5 Controllers for Sampled-Data Systems

Our goal in this section is to design a controller, $K(z)$, for a given sampled-data plant described by its pulse transfer function, $G(z)$. We assume a unity-gain feedback configuration, so that the closed-loop pulse characteristic polynomial, is given as:

$$\Delta(z) = 1 + K(z)G(z)$$

The design of the digital controller for a sampled-data system model can be performed by using the root locus technique. Designing a dynamic controller using $z$-plane root locus can be challenging. Alternately, a controller for a sampled-data system can be obtained by emulating (i.e., approximating) an available continuous-time controller design. These two methods are described next.

### 7.5.1 Root Locus Design of Digital Controllers

The root locus plot in the case of analog systems employs the open-loop transfer function, $KGH(s)$, to describe the locus of the roots of the closed-loop characteristic polynomial: $\Delta(s) = 1 + KGH(s)$, with variation in the controller gain, $K$.

The root locus for the discrete systems is plotted in the $z$-plane. The rules for plotting the $z$-plane locus are similar to those for the $s$-plane locus. We do not need to specify a sampling time for plotting $z$-plane root locus.

The $z$-plane RL describes the locus of roots of the closed-loop pulse characteristic polynomial, $\Delta(z) = 1 + KG(z)$, as controller gain $K$ is varied. A suitable value of $K$ can then be selected form the RL plot. To ensure closed-loop stability, the resulting pulse characteristic polynomial should have its roots inside the unit circle.

In the MATLAB Control Systems Toolbox, "rlocus" command is used to plot both the $s$-plane and $z$-plane root loci. The RL design of a digital controller for sampled-data system is described below.

**Design for a Desired Damping Ratio.** Assuming that the controller design specifications include a desired damping ratio, $\zeta$, for the closed-loop poles, we consider the $z$-plane root locus design based on the $\zeta$ requirement. In particular, we consider a prototype second-order analog transfer function: $T(s) = \frac{\omega_n^2}{s^2 + 2\zeta\omega_n s + \omega_n^2}$, where $s$-plane roots are located at: $s = \sigma \pm j\omega = -\zeta\omega_n \pm \omega_n\sqrt{1 - \zeta^2}$.

The constant $\zeta$ lines in the complex plane are characterized by the relation: $\frac{\sigma}{\omega} = \mp\frac{\zeta}{\sqrt{1-\zeta^2}}$. In the $s$-plane, these are radial lines at angles defined by $\theta = \cos^{-1}\zeta$.

The constant $\zeta$ lines in the $z$-plane are obtained by considering the equivalence: $z = e^{Ts} = e^{\sigma T}e^{\pm j\omega T}$, where $\sigma = \mp\frac{\zeta}{\sqrt{1-\zeta^2}}\omega$. The resulting plot, as $\omega T$ ranges from 0 to $\pi$, gives the constant $\zeta$ lines in the $z$-plane.

For example, let $\zeta = \frac{1}{\sqrt{2}}$; then $\sigma = \mp\omega$, and the constant $\zeta$ line is defined by $z = e^{-\omega T}\angle \pm \omega T$, where $\omega T$ ranges from 0 to $\pi$.

**Damping Ratio Design in MATLAB.** In the MATLAB Control Systems Toolbox, the "grid" command displays the constant $\zeta$ lines in the complex $z$-plane to help in the root locus design of sampled-data systems.

The $z$-plane grid additionally displays constant frequency $(\omega_n/T)$ contours that are helpful in selecting a suitable sampling time $(T)$ given the natural frequency of the continuous-time model.

The desired region for $z$-plane closed-loop root locations may be specified as: $0.1\pi \leq \omega_n T \leq 0.5\pi$, $\zeta \geq 0.6$, which assumes a sampling frequency between 2 and 10 times the natural frequency of the continuous-time system. A sampling frequency higher than $10\omega_n$ can be selected, but requires faster processing and is generally unnecessary.

The root locus design of a sampled-data system is illustrated using the following example.

**Example 7.20:** A second-order sampled-data system (continued from Example 7.18)

Let $G(s) = \frac{1}{s(s+1)}$; we use $T = 0.2$s to obtain: $G(z) = \frac{0.0187z+0.0175}{(z-1)(z-0.819)}$.

For static controller design, the closed-loop pulse characteristic polynomial is given as:

$$\Delta(z) = z^2 + (0.0187\,K - 1.819)z + 0.0175\,K + 0.819.$$

The $z$-plane RL is plotted using MATLAB "rlocus" command. From the RL plot, the controller gain where the $z$-plane RL crosses the $\zeta = 0.7$ line is found as: $K = 0.46$ (Figure 7.8(a)).

The resulting pulse characteristic polynomial is: $\Delta(z) = z^2 - 1.81z + 0.827$, with closed-loop roots located at: $z = 0.9 \pm j0.09$.

We may use the MALTAB "damp" command to check the damping of the closed-loop roots. The results show a damping $\zeta = 0.7$ with a natural frequency $\omega_n = 0.68\ rad/s$.

For comparison, the continuous-time system has a characteristic polynomial: $\Delta(s) = s^2 + s + K$, and for the same controller gain $K = 0.46$, the closed-loop roots are located at $s = 0.5 \pm j0.46$ with $\zeta = 0.74$ with $\omega_n = 0.68\ rad/s$.

The step responses for the continuous-time and discrete-time systems are plotted below (Figure 7.8(b)).

The MATLAB script for this example is given below:

```
G=tf(1,[1 1 0])                 % define plant transfer function
Gz=c2d(G, 0.2)                  % pulse transfer function (T=0.2)
rlocus(Gz),grid                 % plot root locus
Tz=feedback(0.46*Gz,1)          % closed-loop DT system
T=feedback(0.46*G,1)            % closed-loop CT system
damp(Tz)                        % check damping
step(T,Tz),grid                 % plot step response
legend('continuous-time','sampled-data') % figure legend
```

**Design for Settling Time and Damping Ratio.** The settling time and the damping ratio of the dominant closed-loop roots can be deduced

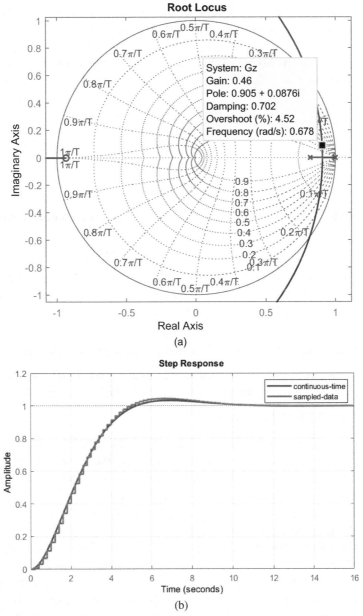

**Figure 7.8**   Root locus design in the *z*-plane (a); step response of the closed-loop system (b).

from the $z$-plane pole locations by making the following comparison based on a prototype second-order system: $re^{\pm j\theta} = e^{-\zeta\omega_n T}e^{\pm j\omega_d T}$, $\omega_d = \omega_n\sqrt{1-\zeta^2}$.

Separating the above expression into real and imaginary part gives two equations: $\ln r = -\zeta\omega_n T$ and $\theta = \omega_d T$, which can be solved to obtain:

$$\zeta = -\ln r/\sqrt{\ln^2 r + \theta^2}$$
$$\omega_n = \sqrt{\ln^2 r + \theta^2}/T$$
$$\tau = \frac{1}{\zeta\omega_n} = -\frac{T}{\ln r}.$$

**Example 7.21:** A second-order sampled-data system (continued from Example 7.20).

Let $G(s) = \frac{1}{s(s+1)}$; we select $T = 0.2s$ to obtain: $G(z) = \frac{0.0187z+0.0175}{(z-1)(z-0.819)}$.

Then, for $K = 0.46$, the $z$-plane closed-loop roots are located at: $z = 0.905 \pm j0.0875 = 0.91\,e^{\pm j0.096}$. Next, we use the above relations to obtain: $\zeta = 0.7$, $\omega_n = 0.74$, $\tau = 2.1\ s$, $t_s = 4.6\tau = 9.5\ s$.

## 7.5.2 Analog and Digital Controller Design Compared

In this section, we intend to compare the analog and digital controller designs using the root locus technique in the case of a DC motor model.

**Example 7.22:** DC motor model.

The analog transfer function of a small DC motor is given as: $G(s) = \frac{500}{s^2+110s+1025}$. We assume that the design specifications for the motor step response are given as: $\%OS \leq 10\%(\zeta \geq 0.59)$, $t_s \leq 100ms$, $e(\infty)_{step} = 0$.

Since the dominant motor time constant is $\tau_m \cong 0.1s$, we initially select a sampling time, $T = 0.02s$; using the MATLAB "c2d" command, the motor pulse transfer function is obtained as: $G(z) = \frac{0.053z+0.026}{z^2-0.95z+0.111}$.

The RL plot for analog system design (Figure 7.9(a)) shows that a range of values, $4 \leq K \leq 15.3$ meets the damping ratio and settling time requirements. We may select, e.g., $K = 10$ for the static controller design.

The root locus plot for the sampled-data system design (Figure 7.9(b)) shows that a range of $K \leq 6$ meets the $\%OS$ requirement; however, the settling time requirement could not be met as $\tau \cong 30ms$ for $K = 6$.

Next, in order to meet the settling time requirement, we reduce the sampling time of $T = 0.01s$, to obtain $G(z) = \frac{0.0178z+0.0123}{z^2-1.27z+0.333}$.

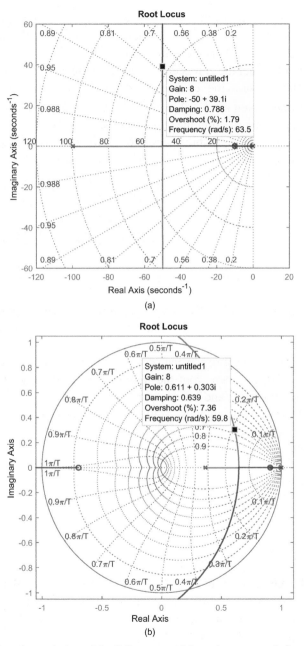

**Figure 7.9** Root locus design of the DC motor model: analog system design (a); sampled-data system design (b).

The static controller that meets the damping ratio requirement in the case of sampled-data system has a modified range: $K \leq 8.55$. We therefore choose $K = 8$ that has $\tau \cong 24$ ms (Figure 7.9(b)).

In order to meet the zero steady-state tracking error requirement, we need to add an integrator to the feedback loop. For the analog plant, we may choose a PI controller given as: $K(s) = \frac{K(s+10)}{s}$, where the controller zero location can be adjusted by plotting the step response.

We define a comparable PI controller for the sampled-data system by using the transformation: $z = e^{Ts}$, which results in: $K(z) = \frac{K(z-0.905)}{z-1}$.

Using $K = 8$ for both analog and sampled-data systems, the resulting closed-loop transfer functions are obtained as:

$$\text{Analog: } T(s) = \frac{4000(s+10)}{s^3 + 110s^2 + 5025s + 40{,}000}$$

$$\text{Discrete: } T(z) = \frac{0.14z^2 - 0.03z - 0.089}{z^3 - 2.13z^2 + 1.57z - 0.42}$$

The use of the MATLAB "damp" command reveal a damping ratio: $\zeta = 0.79$ for the analog system and a damping ratio: $\zeta = 0.68$ for the sampled-data system.

The step responses of the analog and sampled-data systems are compared in Figure 7.10. As seen from the figure, both analog and discrete systems meet the settling time and overshoot requirements.

The MATLAB commands for this example are given below.

```
Gs=tf(500,[1 110 1025])        % define plant transfer function
Gz=c2d(Gs, .01)                % pulse transfer function (T=.1s)
Kpi=tf([1 10],[1 0])           % define PI controller
rlocus(Gs*Kpi)                 % plot root locus with PI (analog)
Kpiz=tf([1 -.905],[1 -1],.01)  % define discrete PI controller
rlocus(Gz*Kpiz)                % plot root locus with PI (discrete)
Ts=feedback(8*Kpi*Gs,1)        % closed-loop system (analog)
Tz=feedback(8*Kpiz*Gz,1)       % closed-loop system (discrete)
damp(Ts), damp(Tz)             % check damping ratio
step(Ts,Tz),grid               % plot step response
legend('analog','discrete')    % figure legend
```

**MATLAB Tuning of PID Controller.** The MATLAB Control System Toolbox provides a utility for tuning of PID controllers. The MATLAB 'pidtune' command can be used with both analog and digital system models. This is explored in the next example.

**Example 7.23:** PID controller for the DC motor model

The simplified model of a small DC motor is given as: $G(s) = \frac{5}{s+10.25}$.

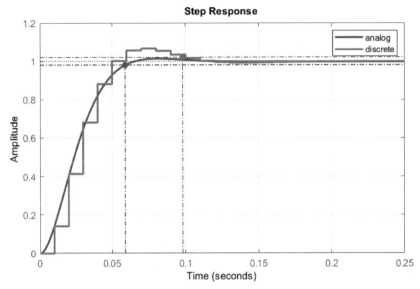

**Figure 7.10** Step response of the DC motor for the analog and discrete controller design.

We use the MATLAB 'pidtune' command to design a PID controller for the DC motor, when a closed-loop bandwidth of 50*rad/s* is specified. The result is: $K(s) = 9.39 + \frac{200}{s}$.

We note that the MATLAB tuned controller has $k_d = 0$.

Next, we use MATLAB 'c2d' command with a sampling time: $T = 0.01s$ to discretize the motor model, which results in: $G(z) = \frac{0.048}{(z-0.9)}$.

The MATLAB 'pidtune' command is used to design a PID controller for the discrete model with similar bandwidth specification. The result is: $K(z) = 12 + 0.072\left(\frac{z-1}{T}\right) + 210\left(\frac{T}{z-1}\right)$.

The unit-step responses of the closed-loop systems for the analog and discrete models are compared (Figure 7.11).

The MATLAB script for this example appears below:

```
Gs=tf(5,[1 10.25]);         % define plant transfer function
Gz=c2d(Gs,.01)              % discretize plant
Kpid=pidtune(Gs,'pid',50)   % design PID controller (analog)
Kzpid=pidtune(Gz,'pid',50)  % design PID controller (discrete)
Ts=feedback(Kpid*Gs,1)      % closed-loop system (analog)
Tz=feedback(Kzpid*Gz,1)     % closed-loop system (discrete)
step(Ts,Tz)                 % plot step response
legend('analog','discrete') % figure legend
```

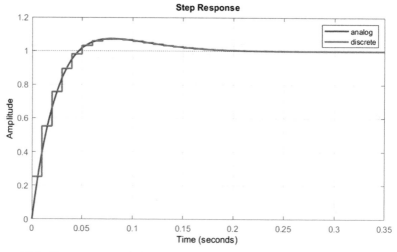

**Figure 7.11**   Step response of the DC motor for the analog and discrete PID controller designs.

### 7.5.3 Digital Controller Design by Emulation

Controller emulation aims to obtain an approximate digital controller, $K(z)$, whose response matches that of the analog controller, $K(s)$. The digital controller can be implemented in a microcontroller or DSP chip using an update rule derived from its time-domain description. Popular controller emulation methods are described below (see references for details).

**Impulse Invariance.** The impulse invariance method aims to match the impulse response of the analog controller. The method involves the following steps and works best for band-limited filters.

1. Obtain the filter transfer function in the partial fraction form: $H(s) = \sum_{k=1}^{n} \frac{A_k}{s - s_k}$
2. Obtain filter impulse response: $h(t) = \sum_{k=1}^{n} A_k e^{s_k t}$
3. Obtain the sampled impulse response: $h(kT) = T \sum_{k=1}^{n} A_k e^{s_k kT}$
4. Apply $z$-transform, to obtain the filter pulse transfer function: $H(z) = T \sum_{k=1}^{n} \frac{A_k}{1 - p_k z^{-1}}, \quad p_k = e^{s_k T}$.

**Pole–Zero Matching.** In pole–zero matching, both the plant poles and zeros are mapped to their equivalent locations in the $z$-plane using: $z_i = e^{s_i T}$. If the analog transfer function has $n$ poles and $m$ finite zeros, to be matched, then another $n - m$ zeros are added at $z = 1$ to the pulse transfer function.

Additionally, the gain of the filter is matched to that of the analog filter at a frequency of interest. For example, the DC gain is matched in the case of a low-pass filter.

**Zero-Order-Hold (ZOH).** The ZOH method assumes the presence of a ZOH that generates a piece-wise constant controller input. The method typically proceeds by realizing the analog controller in the state space, followed by discretization, and conversion to pulse transfer function, $G(z)$. The ZOH method becomes ineffective if the analog controller has poles at the origin (that is, it cannot be used in the case of PI and PID controllers).

**Bilinear Transform (BLT).** The BLT method uses Tustin's approximation $\left(z \cong \frac{1+Ts/2}{1-Ts/2}\right)$ to emulate an analog controller. The BLT is effective with a high sampling frequency, exceeding 10 times the closed-loop bandwidth, that is, for $f_s \geq \frac{10\omega_B}{\pi}$ or $T \leq \frac{\pi}{10\omega_B}$. Using the BLT, the equivalent digital controller is obtained as: $K(z) = K(s)|_{s=\frac{2}{T}\frac{z-1}{z+1}}$. Frequency prewarping can be used to ensure exact matching at a particular frequency (see references).

**Controller emulation in MATLAB.** The MATLAB Control Systems Toolbox provides "c2d" command for discretizing the analog controller that allows the controller emulation method to be specified. The additional choices, besides the above conversion methods, include: first-order-hold (that assumes a piece-wise linear input), and least-squares (to match the frequency response). The default method is ZOH.

Controller emulation methods are illustrated in the following example.

**Example 7.24:** Phase-lead controller design.

A phase-lead controller for the transient response improvement of the analog plant, described by: $G(s) = \frac{2}{s(s+1)(s+2)}$ was previously designed as: $K(s) = \frac{14.58(s+0.36)}{s+5.25}$. Let $(T = 0.1s)$; then, the resulting digital controllers using the above controller emulation techniques are given below.

**Impulse Invariance.** Since $K(s) = 14.58\left(1-\frac{4.89}{s+5.25}\right)$, the equivalent digital controller is obtained as: $K(z) = 14.58\left(1 - \frac{0.489}{z-0.592}\right) = \frac{7.45(z-0.957)}{z-0.592}$. The update rule for the digital controller implementation in a computer is given as: $u_k = 0.592u_{k-1} + 7.45(e_k - 0.957e_{k-1})$.

**Pole–Zero Matching.** An equivalent digital controller obtained by pole–zero matching $(T = 0.1s)$ is given as: $K(z) = \frac{K(z+e^{0.36T})}{z+e^{5.25T}}$. If the DC gain is matched, the equivalent controller is given as: $K(z) = \frac{11.548(z-0.965)}{z-0.592}$. Alternatively, we may want to match the filter gain at $\omega_m = \sqrt{\omega_z\omega_p}$, which gives:

$K(z) = \frac{5.177(z-0.965)}{z-0.592}$. The update rule for digital controller implementation on a computer is given as: $u_k = 0.592u_{k-1} + 5.177(e_k - 0.965e_{k-1})$.

**Zero-Order-Hold (ZOH).** A state-space model of the analog controller is obtained as: $\dot{x} = -5.25x + 8.845e$, $\quad u = -8.06x + 14.58e$. An equivalent discrete-time model $(T = 0.1s)$ is given as: $x_{k+1} = 0.592x_k + 0.688e_k$, $\quad u_k = -8.06x_k + 14.58u_k$. The corresponding pulse transfer function is given as: $K(z) = \frac{14.58\,(z-0.972)}{z-0.592}$. The update rule for digital controller implementation on a computer is given as: $u_k = 0.592u_{k-1} + 14.58(e_k - 0.972e_{k-1})$.

**Bilinear Transform (BLT).** The analog compensated system has a bandwidth: $\omega_B \cong 1.8\ \frac{rad}{s}$. Using $T = 0.1s$, a matching digital controller is obtained as: $K(z) = \frac{11.756(z-0.965)}{z-0.584}$. The update rule for digital controller implementation on a computer is given as: $u_k = 0.584u_{k-1} + 11.56(e_k - 0.965e_{k-1})$.

**Comparison of Controller Emulation Methods.** Figure 7.12 compares the step response of the closed-loop system using impulse invariance, pole–zero matching, bilinear transform, and the ZOH emulation methods.

The MATLAB script used for the above example is given below:

```
Gs=tf(2, [1 3 2 0]);            % define plant transfer function
Ks=zpk(-.36,-5.25,14.58);       % controller transfer function
Gz=c2d(G,.1);                   % discretize plant tranfer function
Kzi=zpk(.957,.592,7.45,-1);     % controller by impulse invariance
Kzm=c2d(Ks,.1, 'matched');      % controller by pz matching
Kzt=c2d(Ks,.1, 'tustin');       % controller by BLT
Kzz=c2d(Ks,.1, 'zoh');          % controller by ZOH
Ts=feedback(Ks*Gs,1);           % closed-loop system (analog)
Tzi=feedback(Kzi*Gz,1);         % closed-loop system (impulse)
Tzm=feedback(Kzm*Gz,1);         % closed-loop system (matching)
Tzt=feedback(Kzt*Gz,1);         % closed-loop system (Tustin's)
Tzz=feedback(Kzz*Gz,1);         % closed-loop system (ZOH)
step(Ts,Tzi,Tzm,Tzt,Tzz),grid  % plot step response
legend('analog', 'impulse', 'pzmatch', 'tustin', 'zoh')
```

As seen from the plots, the bilinear transform and pole–zero matching methods produce comparable step responses with about 20% overshoot, compared to about 28% overshoot for the ZOH method. The impulse invariance method has a much lower 4% overshoot. For comparison, the continuous-time closed-loop system has about 15% overshoot.

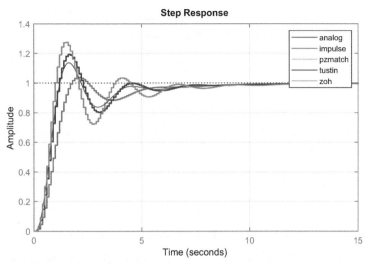

**Figure 7.12** Step response comparison for controller emulation methods: MATLAB simulation using the pulse transfer function (a); Simulink simulation with the analog plant (b).

## 7.5.4 Emulation of Analog PID Controller

The PID controller is popular in industrial process control applications due to its versatility and robustness against process model mismatches. The industrial PID controller is commonly implemented on an industrial computer or a programmable logic controller (PLC).

An analog PID controller with input $e(t)$ and output $u(t)$ is described as:

$$u(t) = k_p e(t) + k_d \frac{de(t)}{dt} + k_i \int e(t)\,dt.$$

An equivalent digital PID controller can be obtained by employing a forward difference or backward difference approximation of the derivative term and its inverse for the integral term, that is,

For the differentiator, use $u_k = \frac{1}{T}(e_k - e_{k-1})$ or $u_k = \frac{1}{T}(e_{k+1} - e_k)$

For the integrator, use $u_k = u_{k-1} + Te_k$ or $u_k = u_{k-1} + Te_{k-1}$.

The resulting $z$-domain expressions for the PID controller are given as:

Forward Euler: $K(z) = k_p + k_d \left( \dfrac{z-1}{T} \right) + k_i \left( \dfrac{T}{z-1} \right)$

Backward Euler: $K(z) = k_p + k_d \left( \dfrac{z-1}{Tz} \right) + k_i \left( \dfrac{Tz}{z-1} \right)$.

Let $v_k$ denote the integrator output; then, Backward Euler PID controller is implemented using the following update rules:

$$u_k = k_p e_k + \frac{k_d}{T}(e_k - e_{k-1}) + k_i v_k$$

$$v_k = v_{k-1} + T e_k.$$

Whereas, the Forward Euler PID controller is implemented as:

$$u_k = k_p e_k + \frac{k_d}{T}(e_{k+1} - e_k) + k_i v_k$$

$$v_k = v_{k-1} + T e_{k-1}$$

**Example 7.25:** Digital emulation of analog PID controller.

An analog PID controller for the analog plant: $G(s) = \frac{1}{s(s+2)(s+5)}$ was earlier designed as: $K(s) = \frac{12.5(s+0.05)(s+2.05)}{s} = 26.25 + 12.5s + \frac{1.28}{s}$. Using the above emulation technique, an equivalent digital PID controller is given as: $K(z) = 26.25 + 12.5\left(\frac{z-1}{Tz}\right) + 1.28\left(\frac{Tz}{z-1}\right)$.

In particular, for $T = 0.1s$, the PID controller emulation gives: $K(z) = \frac{151.378(z-0.995)(z-0.83)}{z(z-1)}$.

The update rules for the computer implementation of the digital PID controller are given as:

$$u_k = 26.25 e_k + 125(e_k - e_{k-1}) + 1.28 v_k$$

$$v_k = v_{k-1} + 0.1 e_k.$$

For comparison, equivalent digital PID controllers were obtained using the following methods.

$$\text{Pole-zero matching: } K(z) = \frac{138.52(z - 0.995)(z - 0.815)}{(z - 1)}.$$

$$\text{Bilinear transform: } K(z) = \frac{276.31(z - 0.995)(z - 0.814)}{(z - 1)(z + 1)}.$$

The following plot (Figure 7.13) shows a comparison of the step responses for the three PID controllers. The pulse transfer function for the analog plant was obtained from the ZOH method. From the figure, the pole–zero matching controller provides the best PID controller emulation

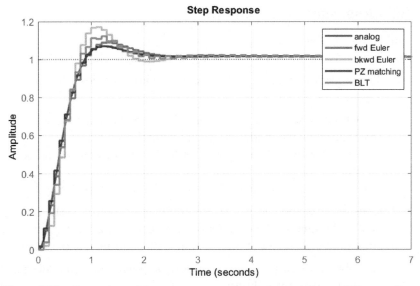

**Figure 7.13**   Comparison of the step response using analog and discrete PID controllers.

for this example. Further, forward Euler approximation performs better than backward Euler.

The MATLAB script used for this example is given below:

```
s=tf('s')                      % define 's'
Gs=zpk([],[0 -2 -5],1);        % define plant transfer function
Ks=26.25+12.5*s+1.28/s;        % controller transfer function
Gz=c2d(Gs,.1)                  % discretize plant transfer function
Kz1=26.25+12.5/.1*tf([1 -1],1,.1)+1.28*.1/tf([1 -1],1,.1)
                               % PID (Fwd Euler)
Kz2=26.25+12.5/.1*tf([1 -1],[1 0],.1)+1.28*.1/tf([1 -1],[1 0],.1)
                               % PID (Bkwd Euler)
Kz3=c2d(Ks,.1, 'matched');     % PID controller (PZ matching)
Kz4=c2d(Ks,.1, 'tustin');      % PID controller (BLT)
Ts=feedback(Ks*Gs,1);          % closed-loop system (CT)
Tz1=feedback(Kz1*Gz,1);        % closed-loop system (FWD Euler)
Tz2=feedback(Kz2*Gz,1);        % closed-loop system (Bkwd Euler)
Tz3=feedback(Kz3*Gz,1);        % closed-loop system (PZ Matching)
Tz4=feedback(Kz4*Gz,1);        % closed-loop system (BLT)
step(Ts,Tz1,Tz2,Tz3,Tz4),grid  % plot step response
legend('anlaog','fwd Euler','bkwd Euler','matched','tustin')
                               % figure legend
```

## Skill Assessment Questions

Link to the answers:

http://www.riverpublishers.com/book_details.php?book_id=449

1. The model of an environmental control system is described as: $G(s) = \frac{1}{10s+1}$.

    (a) Choose a sampling time $T$ and discretize the plant model.
    (b) Determine the modes of system natural response.
    (c) Obtain a time-domain description for the discretized model of the plant.
    (d) Solve for the first few terms of the unit-step response by iteration.

2. Consider the model of the environmental control system in Question 1. Assume that the system is connected in a unity-gain feedback configuration with a sampler, a static gain controller, and a ZOH.

    (a) Obtain the closed-loop characteristic polynomial and determine the range of $K$ for stability.
    (b) Use $z$-plane root locus to design a controller for the system so that the closed-loop unit-step response settles in $2s$.
    (c) Plot the unit-step response of the closed-loop system to verify the design.

3. The simplified model of a DC motor with position output is given as: $G(s) = \frac{5}{s+10}$.

    (a) Choose a sampling time $T$ and discretize the plant.
    (b) Determine the modes of system natural response.
    (c) Obtain a time-domain description for the discretized model of the plant.
    (d) Solve for the first few terms of the unit-pulse response by iteration.

4. Consider the model of the DC motor in Question 3. Assume that the system is connected in a unity-gain feedback configuration with a sampler, a static gain controller, and a ZOH.

    (a) Obtain the closed-loop characteristic polynomial and determine the range of $K$ for stability.
    (b) Use $z$-plane root locus to design a controller for the system so that the closed-loop system has $\zeta \cong 0.65$.
    (c) Plot the unit-step response of the closed-loop system to verify the design.

5. A robot vision control system is modeled as: $G(s) = \frac{2}{(s+1)(s+2)}$.

   (a) Choose a sampling time $T$ and discretize the plant.
   (b) Determine the modes of system natural response.
   (c) Obtain a time-domain description for the discretized model of the plant.
   (d) Solve for the first few terms of the unit-pulse response by iteration.

6. Consider the discretized model of robot vision control system in Question 5. Assume that the system is connected in a unity-gain feedback configuration with a sampler, a controller, and a ZOH.

   (a) Use MATLAB 'pidtune' command to design PI controllers for the analog and discrete models.
   (b) Plot and compare the unit-step response of the closed-loop systems.

7. The model of the pitch axis of a quad-copter with actuator is described as: $G(s) = \frac{1}{s^2(s+10)}$. It is desired that the command response should settle in less than 3 seconds.

   (a) Choose a sampling time and obtain a pulse transfer function for the plant.
   (b) Choose a bandwidth and use MATLAB 'pidtune' command to design a PID controller for the analog plant model.
   (c) Using similar bandwidth with MATLAB 'pidtune' command to design a PID controller for the discrete plant model.
   (d) Compare the unit-step responses of the closed-loop analog and discrete systems.

8. Consider the simplified model of a flexible structure given as: $G(s) = \frac{0.5}{s} + \frac{0.1}{s^2+s+100}$.

   (a) Obtain a pulse transfer function for the plant (use $T = 0.01s$).
   (b) Choose a bandwidth and use MATLAB 'pidtune' command to design a PID controller for the analog plant model.
   (c) Using similar bandwidth with MATLAB 'pidtune' command to design a PID controller for the discrete plant model.
   (d) Compare the unit-step responses of the closed-loop analog and discrete systems.

9. The rate loop for a certain missile autopilot is modeled as: $G(s) = \frac{100(s+1)}{s^2+2s+100}$. Assume that a PI controller for the autopilot is designed as: $K(s) = 0.2 + \frac{5}{s}$.

(a) Choose a sampling time and obtain the pulse transfer function for the plant.

(b) Emulate the controller using BLT and PZ matching techniques. Plot and compare the step responses of the closed-loop systems.

10. The model of a satellite position control system is given as: $G(s) = \frac{1}{s^2(s+50)}$. Assume that a lead-lag controller for the satellite model is designed as: $K(s) = \frac{1 \times 10^5 (s+1)(s+10)}{(s+0.1)(s+100)}$.

(a) Choose a sampling time and obtain the pulse transfer function for the plant.

(b) Emulate the controller using the BLT method.

(c) Plot and compare the step responses of the closed-loop systems for the analog and discrete models.

(d) Give an update rule for controller implementation on computer.

11. Consider the model of the satellite position control system in Question 10. Assume that a PID controller for the satellite model is designed as: $K(s) = 750(2 + s + \frac{1}{s})$.

(a) Choose a sampling time and obtain the pulse transfer function for the plant.

(b) Emulate the PID controller using forward difference method.

(c) Plot and compare the step response of the closed-loop system for the analog and discrete models.

(d) Give an update rule for controller implementation on computer.

12. The simplified model of an automobile given as: $G(s) = \frac{28s+120}{s^2+7s+14}$. Assume that the system is connected in a unity-gain feedback configuration with a sampler, static controller, and a ZOH.

(a) Choose a sampling time $T$ and obtain a pulse transfer function for the plant.

(b) Determine the range of $K$ for the stability of the closed-loop characteristic polynomial. Choose $K = \frac{1}{2}K_{max}$, where $K_{max}$ denotes the maximum value for closed-loop stability.

(c) Determine the steady-state error to a step reference input.

(d) Plot and compare the step responses of the closed-loop analog and discrete systems.

13. Consider a system model described as: $G(s) = \frac{s+3}{s(s+1)(s+2)}$. Assume that the system is connected in a unity-gain feedback configuration with a sampler, static controller, and a ZOH.

   (a) Choose a sampling time $T$ and obtain the pulse transfer function for the plant.
   (b) Determine the range of $K$ for the stability of the closed-loop characteristic polynomial. Choose $K = \frac{1}{2}K_{max}$, where $K_{max}$ denotes the maximum value for closed-loop stability.
   (c) Determine the steady-state error to a ramp reference input.
   (d) Plot and compare the ramp responses of the closed-loop analog and discrete systems.

14. Consider the model of human postural dynamics described as an inverted pendulum, given as: $G(s) = \frac{k}{s^2-\Omega^2}$, where $k = 0.01$, $\Omega = \sqrt{10}$.

   (a) Obtain a pulse transfer function for the model (use $T = 0.1s$).
   (b) Obtain the first few terms of the unit-pulse response.
   (c) Use MATLAB 'pidtune' command to design PID controllers for analog and discrete models.
   (d) Plot and compare the unit-step responses for the closed-loop analog and discrete systems.
   (e) Give an update rule for the implementation of the controller.

# 8

# Controller Design for State Variable Models

## Learning Objectives

1. Perform state feedback controller design for state variable system models.
2. Design pole placement and reference tracking controllers for state variable models.
3. Obtain and a analyze sampled-data system models in state variable form.
4. Perform digital controller design for sampled-data systems in state variable form.

The state variable model of a dynamic system comprises first-order differential equations that express the time derivatives of a set of state variables, selected to adequately describe system behavior. The state variables are often selected as the natural variables associated with the energy storage elements present in the system. Examples of such variables are the capacitor voltages and inductor currents in electrical circuits, and the displacement and velocity of the inertial elements in mechanical systems.

The state feedback controller design for a given state variable model involves selecting feedback gains for all (or a selection of) the state variables. The state feedback controller design offers greater flexibility in control system design compared to designing output feedback controllers with transfer function models (discussed in Chapter 6 earlier).

The pole placement design using full-state feedback refers to the selection of $n$ feedback gains for placing the $n$ roots of the closed-loop characteristic polynomial at desired locations in the complex plane. The pole placement design is facilitated when the state variable model is in controller form. Alternately, Ackermann's and Bass–Gura formulas, or Sylvester's equation can be used for designing the state feedback controller.

The tracking system design involves reducing steady-state error to a given reference input to zero. This can be achieved by feeding forward the reference signal to cancel the tracking error; however, the design is not robust against parameter variations. A more robust design involves integrating the error signal inside the feedback loop, and can be realized by performing full-state feedback on an augmented state variable model that includes the differential equation describing output of the integrator.

The state variable design methods can be extended to digital controller design of sampled-data systems in state variable form. The discrete-time state variable description of a system is obtained by discretizing the continuous-time state equations with a zero-order-hold (ZOH), and is applicable at the sampling instants. A solution to the discrete-state equations is easily obtained by iteration.

The pole placement design using state feedback is similarly performed on discrete state variable models. The desired characteristic polynomial in the discrete case has its roots inside the unit circle in order to ensure stability of the closed-loop system. A peculiarity in the case of discrete system models is the design of a deadbeat controller that places all roots of the closed-loop characteristic polynomial at the origin. The deadbeat controller ensures that the system response reaches the steady state in exactly $n$ iterations.

Compared to the root locus design that allows selective placement of the closed-loop poles of the characteristic polynomial, arbitrary pole placement is possible through the use of state feedback design in the case of state variable models. In the following, we assume that the system to be controlled is of single-input single-output (SISO) type and has $n$ state variables.

## 8.1 State Feedback Controller Design

The state feedback controller refers to feeding back the state variables and using that information to control the plant input as a means to steer the plant output. The overall design goal remains to cause the output to meet the specified design criteria.

Let $\mathbf{x}(t)$ denote a vector of $n$ state variables, $u(t)$ denote a scalar input, and $y(t)$ denote a scalar output; then the state variable model of a SISO system is written as:

$$\dot{\mathbf{x}}(t) = \mathbf{A}\mathbf{x}(t) + \mathbf{b}u(t)$$
$$y(t) = \mathbf{c}^T\mathbf{x}(t).$$

In the above, $\mathbf{A}$ is a $n \times n$ system matrix, $\mathbf{b}$ is a $n \times 1$ column vector that distributes the input $u(t)$, and $\mathbf{c}^T$ is a $1 \times n$ row vector that describes individual state variable contributions toward the output $y(t)$. The state variable model, as stated above, represents an input-output transfer function that is strictly proper.

In a more general case of multi-input multi-output systems, we may assume that the system model has an input vector $u(t)$, a distribution matrix $B$, an output collection matrix $C$, and an output vector $y(t)$, to express the resulting state variable model as:

$$\dot{\mathbf{x}}(t) = \mathbf{A}\mathbf{x}(t) + \mathbf{B}u(t)$$
$$\mathbf{y}(t) = \mathbf{C}\mathbf{x}(t).$$

In the following, we discuss the popular pole placement controller design method.

### 8.1.1 Pole Placement with State Feedback

The state feedback design involves feeding back all the state variables to generate the error signal. An underlying assumption here is that all the state variables are available for measurement. If this not the case, the missing state variables must be reconstructed using a state estimator (see references).

Assuming that all the state variables are available for feedback, the control law is expressed as:

$$u = -\mathbf{k}^T\mathbf{x} + r,$$

where $\mathbf{k}^T = [k_1, k_2, \ldots, k_n]$ is a vector of $n$ feedback gains to be selected, one for each of the $n$ state variables, and $r$ is a scalar reference input, which may be constant as in the case of output regulation, or time varying as in the case of reference tracking.

By including state feedback in the state equations, the closed-loop system dynamics are described as:

$$\dot{\mathbf{x}}(t) = (\mathbf{A} - \mathbf{b}\mathbf{k}^T)\mathbf{x}(t) + \mathbf{b}r.$$

The design problem, then, is to select the feedback gains $\mathbf{k}$ to appropriately scale the state variable vector $\mathbf{x}(t)$, such that the closed-loop system matrix $\mathbf{A} - \mathbf{b}\mathbf{k}^T$ has a characteristic polynomial that aligns with a desired polynomial with suitable root locations. This condition is expressed as:

$$|s\mathbf{I} - \mathbf{A} + \mathbf{b}\mathbf{k}^T| = \Delta_{des}(s)$$

The above equation represents an $n$th order polynomial equation; therefore, by equating the coefficients on both sides of the equation, we obtain $n$ linear equations that can be solved for the $n$ feedback gains: $k_i, i = 1, \ldots, n$.

The desired characteristic polynomial for pole placement design is selected with root locations that meet the time-domain design specifications, as illustrated in the following example.

**Example 8.1:** Pole placement design of a DC motor

The state and output equations for a small DC motor model are given as:

$$\frac{d}{dt}\begin{bmatrix} i_a \\ \omega \end{bmatrix} = \begin{bmatrix} -100 & -5 \\ 5 & -10 \end{bmatrix}\begin{bmatrix} i_a \\ \omega \end{bmatrix} + \begin{bmatrix} 100 \\ 0 \end{bmatrix} V_a$$

$$\omega = \begin{bmatrix} 0 & 1 \end{bmatrix}\begin{bmatrix} i_a \\ \omega \end{bmatrix}.$$

The DC motor transfer function has a denominator polynomial $d(s) = s^2 + 110s + 1025$, with roots located at $s_{1,2} = -10.28, -99.72$; these roots correspond to the motor time constants ($\tau_e \cong 10$ m sec, $\tau_m \cong 100$ m sec).

Assume that we wish to improve the motor transient response by reducing the slower of the time constants from 100 msec to 20 msec. This can be accomplished by using pole placement design to move the dominant closed-loop pole to $-50$; the other pole location is selected at $-100$ (it can also be left at its current location, $-99.72$).

For the selected root locations, the desired characteristic polynomial is given as: $\Delta_{des}(s) = s^2 + 150s + 5000$.

Since the reference signal does not affect pole placement, we may assume $r = 0$, so that the control law is given as: $V_a = -\mathbf{k}^T\mathbf{x}$, where $\mathbf{k}^T = [k_1, k_2]$.

With the introduction of state feedback, the closed-loop system model is given as:

$$\mathbf{A} - \mathbf{bk}^T = \begin{bmatrix} -100(1 + k_1) & -5(1 + 20k_2) \\ 5 & -10 \end{bmatrix}$$

The closed-loop characteristic polynomial is obtained as:

$$|s\mathbf{I} - \mathbf{A} + \mathbf{bk}^T| = (s + 10)[s + 100(1 + k_1)] + 25(1 + 20k_2).$$

The coefficients of the above characteristic polynomial are functions of controller gains. Hence, by comparing the coefficients of the closed-loop characteristic polynomial with those of $\Delta_{des}(s)$, we obtain the following feedback gains: $k_1 = 0.4, k_2 = 7.15$. The resulting control law is given as:

$$V_a(t) = -0.4\, i_a(t) - 7.15\, \omega(t).$$

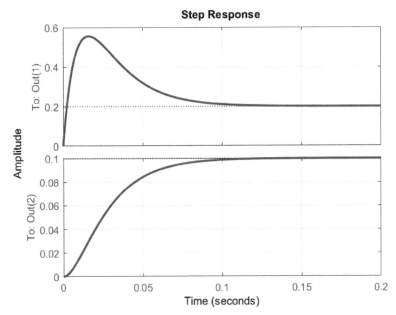

**Figure 8.1** The step response of the DC motor with pole placement controller: armature current (top); motor speed (bottom).

The step response of the DC motor for the above controller design is plotted in Figure 8.1.

The MATLAB commands for plotting the step response are given below.

```
A=[-100 -5;5 -10]; B=[100;0]; % define state variable model
K=[0.4 7.15];                  % pole placement controller
step(A-B*K,B,eye(2),[0;0]), grid % plot step response
```

A necessary condition for pole placement using state feedback is that the underlying system is controllable (Section 3.3.1). A system that is controllable can be transformed into the controller form, which facilitates the pole placement design, as described next.

## 8.1.2 Pole Placement in the Controller Form

The controller design for the state variable system models is relatively easier if the system and input matrices are in controller form (Section 3.5). The state variables in the controller form realization include the output and its derivatives. The controller form assumes a structure where the coefficients of the characteristic polynomial appear in reverse order in the last row of **A** matrix. The controller form structure is given as:

$$\mathbf{A} = \begin{bmatrix} 0 & 1 & 0 & \cdots \\ 0 & 0 & 1 & \cdots \\ \vdots & \vdots & \ddots & 1 \\ -a_n & -a_{n-1} & \cdots & -a_1 \end{bmatrix}, \quad \mathbf{b} = \begin{bmatrix} 0 \\ 0 \\ \vdots \\ 1 \end{bmatrix}.$$

Then, using state feedback, $u = -\mathbf{k}^T\mathbf{x} + r$, the closed-loop system matrix is given as:

$$\mathbf{A} - \mathbf{bk}^T = \begin{bmatrix} 0 & 1 & 0 & \cdots \\ 0 & 0 & 1 & \cdots \\ \vdots & \vdots & \ddots & 1 \\ -a_n - k_1 & -a_{n-1} - k_{n-1} & \cdots & -a_1 - k_n \end{bmatrix}.$$

The closed-loop characteristic polynomial can be written by inspection; the polynomial includes the controller gains, $k_i$, $i = 1, \ldots n$, and is given as:

$$\Delta(s) = s^n + (a_1 + k_n)s^{n-1} + \cdots + a_n + k_1$$

Next, let the desired characteristic polynomial be defined as:

$$\Delta_{des}(s) = s^n + \bar{a}_1 s^{n-1} + \cdots + \bar{a}_{n-1}s + \bar{a}_n.$$

The feedback gains, obtained by comparing the polynomial coefficients, are:

$$k_1 = \bar{a}_n - a_n, \quad k_2 = \bar{a}_{n-1} - a_{n-1}, \ldots, \quad k_n = \bar{a}_1 - a_1.$$

Since the state variables in the controller form include the output and its derivatives, pole placement using state feedback amounts to a generalization of the rate feedback controller (Section 4.2.3), or the proportional-derivative (PD) controller.

**Example 8.2:** The mass–spring–damper model.

Consider the mass–spring–damper model with the mass position and velocity variables as state variables, where the following parameter values are assumed: $m = 1, b = 1, k = 10$. The resulting state variable model is in controller form and is given as:

$$\frac{d}{dt}\begin{bmatrix} x \\ v \end{bmatrix} = \begin{bmatrix} 0 & 1 \\ -10 & -1 \end{bmatrix}\begin{bmatrix} x \\ v \end{bmatrix} + \begin{bmatrix} 0 \\ 1 \end{bmatrix}fx = \begin{bmatrix} 1 & 0 \end{bmatrix}\begin{bmatrix} x \\ v \end{bmatrix}.$$

The characteristic polynomial of the mass–spring–damper system can be written by inspection as the system is already in the controller form and is given as: $\Delta(s) = s^2 + s + 10$.

The system has low damping, and, in order to improve the damping, the desired characteristic polynomial is selected as: $\Delta_{des}(s) = s^2 + 4s + 10$.

The desired feedback gains can be written by inspection, and are given as: $\mathbf{k}^T = [0 \ 3]$. The resulting state feedback controller is described as: $f = -3v$

### 8.1.3 Pole Placement using Bass–Gura Formula

The Bass–Gura formula describes a simple expression to compute the feedback gains from the coefficients of the available and desired closed-loop characteristic polynomials. The formula is derived as follows. Let the state variable model be given as:

$$\dot{\mathbf{x}}(t) = \mathbf{A}\mathbf{x}(t) + \mathbf{b}u(t), \quad y(t) = \mathbf{c}^T \mathbf{x}(t)$$

Assuming that the state variable model is controllable, it can be transformed into controller form by a linear transformation, $\mathbf{z} = \mathbf{P}\mathbf{x}$. The resulting controller form is described as:

$$\dot{\mathbf{z}}(t) = \mathbf{A}_{CF}\mathbf{z}(t) + \mathbf{b}_{CF}u(t), \quad y(t) = \mathbf{c}_{CF}^T \mathbf{z}(t)$$

The controllability matrix of the given model is formed as: $\mathbf{M}_C = [\mathbf{b}, \mathbf{A}\mathbf{b}, \ldots, \mathbf{A}^{n-1}\mathbf{b}]$. Further, the controllability matrix of the controller form representation is given as: $\mathbf{M}_{CF} = [\mathbf{b}_{CF}, \mathbf{A}_{CF}\mathbf{b}_{CF}, \ldots, \mathbf{A}_{CF}^{n-1}\mathbf{b}_{CF}]$.

The matrix that transforms the given state-space model into its controller form representation is defined by:

$$\mathbf{P}^{-1} = \mathbf{M}_C \mathbf{M}_{CF}^{-1}, \quad \mathbf{P} = \mathbf{M}_{CF} \mathbf{M}_C^{-1}.$$

Having transformed the state variable model into the controller form, the full-state feedback control law is defined as:

$$u = -\mathbf{k}_{CF}^T \mathbf{z}(t) = -\mathbf{k}_{CF}^T \mathbf{P}\mathbf{x}(t).$$

where $\mathbf{k}_{CF}^T = [\bar{a}_n - a_n \quad \bar{a}_{n-1} - a_{n-1} \cdots \bar{a}_1 - a_1]$.

The feedback gains for the original state variable model are obtained as: $\mathbf{k}^T = \mathbf{k}_{CF}^T \mathbf{P}$. Hence, the Bass–Gura formula is given as:

$$\mathbf{k}^T = [\bar{a}_n - a_n \quad \bar{a}_{n-1} - a_{n-1} \cdots \bar{a}_1 - a_1] \ \mathbf{M}_{CF} \mathbf{M}_C^{-1}$$

We note that for a given state variable model, the $\mathbf{M}_{CF}$ matrix includes the coefficients of the characteristic polynomial, $\Delta(s) = |sI - A|$; moreover, it can be written by inspection for simple models, as illustrated in the following example.

**Example 8.3:** The DC motor model.

The state equation for a small DC motor model is given as:

$$\frac{d}{dt}\begin{bmatrix} i_a \\ \omega \end{bmatrix} = \begin{bmatrix} -100 & -5 \\ 5 & -10 \end{bmatrix}\begin{bmatrix} i_a \\ \omega \end{bmatrix} + \begin{bmatrix} 100 \\ 0 \end{bmatrix} V_a$$

The controllability matrix for the model is given as: $\mathbf{M}_C = [\mathbf{b}, \ \mathbf{Ab}] = \begin{bmatrix} 100 & -10^4 \\ 0 & 500 \end{bmatrix}$.

The characteristic polynomial of the model is: $|s\mathbf{I} - \mathbf{A}| = s^2 + 110s + 1025$. Hence, its controller form representation is developed as:

$$\frac{d}{dt}\begin{bmatrix} x_1 \\ x_2 \end{bmatrix} = \begin{bmatrix} 0 & 1 \\ -1025 & -110 \end{bmatrix}\begin{bmatrix} x_1 \\ x_2 \end{bmatrix} + \begin{bmatrix} 0 \\ 1 \end{bmatrix} V_a$$

Next, the controllability matrix for the controller form representation is: $\mathbf{M}_{CF} = \begin{bmatrix} 0 & 1 \\ 1 & -110 \end{bmatrix}$.

Let the desired characteristic polynomial be given as: $\Delta_{des}(s) = s^2 + 150s + 5000$.

The feedback gains for the controller form representation are obtained as: $\mathbf{k}_{CF}^T = \begin{bmatrix} 3975 & 40 \end{bmatrix}$.

Using the Bass–Gura formula, the state feedback controller gains for the given DC motor model are computed as: $\mathbf{k}^T = \mathbf{k}_{CF}^T \mathbf{M}_{CF}\mathbf{M}_C^{-1} = \begin{bmatrix} 0.4 & 7.15 \end{bmatrix}$.

The MATLAB commands for this example are given below:

```
A=[-100 -5;5 -10]; B=[100;0];  % define state variable model
pA=poly(A);                     % characteristic polynomial
M=[B, A*B];                     % define controllability matrix
Mc=[0 1;1 -pA(2)];              % C-form controllability matrix
Pd=[1 150 5000];                % desired characteristic polynomial
Kc= fliplr(pd(2:3)-pA(2:3));    % C-form controller gains
K= Kc*Mc/M                      % controller gains for the model
```

## 8.1.4 Pole Placement using Ackermann's Formula

The Ackermann's formula is, likewise, a simple expression to compute the full-state feedback controller gains for a given state variable model, so that the closed-loop eigenvalues match with the roots of a desired characteristic polynomial. To develop the formula, we assume an $n$-dimensional state variable model, given as:

$$\dot{\mathbf{x}}(t) = \mathbf{A}\mathbf{x}(t) + \mathbf{b}u(t), \quad y(t) = \mathbf{c}^T\mathbf{x}(t)$$

The controllability matrix of the given state variable model is formed as: $\mathbf{M}_C = [\mathbf{b}, \mathbf{A}\mathbf{b}, \ldots, \mathbf{A}^{n-1}\mathbf{b}]$.

We also assume that a desired characteristic polynomial is given as: $\Delta_{des}(s) = s^n + \bar{a}_1 s^{n-1} + \cdots + \bar{a}_{n-1}s + \bar{a}_n$. Next, we evaluate the following matrix polynomial that involves a polynomial function of the system matrix:

$$\Delta_{des}(\mathbf{A}) = \mathbf{A}^n + \bar{a}_1\mathbf{A}^{n-1} + \cdots + \bar{a}_{n-1}\mathbf{A} + \bar{a}_n\mathbf{I}.$$

In the above, $\mathbf{I}$ denotes an $n \times n$ identity matrix. The state feedback controller gains are computed from the following formula (known as the Ackermann's formula):

$$\mathbf{k}^T = \begin{bmatrix} 0 \cdots 0 & 1 \end{bmatrix} \mathbf{M}_C^{-1}\Delta_{des}(\mathbf{A})$$

The application of the Ackermann formula is illustrated by the following examples.

**Example 8.4:** The DC motor model.

The state equations for a small DC motor model are given as:

$$\frac{d}{dt}\begin{bmatrix} i_a \\ \omega \end{bmatrix} = \begin{bmatrix} -100 & -5 \\ 5 & -10 \end{bmatrix}\begin{bmatrix} i_a \\ \omega \end{bmatrix} + \begin{bmatrix} 100 \\ 0 \end{bmatrix}V_a$$

Let the desired characteristic polynomial be given as: $\Delta_{des}(s) = s^2 + 150s + 5000$.

Next, the matrix polynomial is evaluated as: $\Delta_{des}(\mathbf{A}) = \begin{bmatrix} -25 & -200 \\ 200 & 3575 \end{bmatrix}$.

The controllability matrix for the model is computed as: $\mathbf{M}_C = [\mathbf{b}, \mathbf{A}\mathbf{b}] = \begin{bmatrix} 100 & -10^4 \\ 0 & 500 \end{bmatrix}$.

Using the Ackermann formula, the pole placement controller gains are computed as: $k^T = \begin{bmatrix} 0.4 & 7.15 \end{bmatrix}$. The resulting state feedback control law is given as: $V_a = -0.4i_a - 7.15\omega$.

The MATLAB commands for this example are given below:

```
A=[-100 -5;5 -10]; B=[100;0]; % define state variable model
Mc=ctrb(A,B);                 % define controllability matrix
I=eye(size(A));               % define identity matrix
pd=[1 150 5000];              % desired characteristic polynomial
K=I(end,:)/Mc*polyvalm(pd,A); % controller gains
```

**Example 8.5:** The mass–spring–damper model.

The state equation for the mass–spring–damper model is described as:

$$\frac{d}{dt}\begin{bmatrix} x \\ v \end{bmatrix} = \begin{bmatrix} 0 & 1 \\ -10 & -1 \end{bmatrix} \begin{bmatrix} x \\ v \end{bmatrix} + \begin{bmatrix} 0 \\ 1 \end{bmatrix} f$$

Let a desired characteristic polynomial be selected as: $\Delta_{des}(s) = s^2 + 4s + 10$;

Then, the matrix polynomial is evaluated as: $\Delta_{des}(A) = \begin{bmatrix} 0 & 3 \\ -30 & -3 \end{bmatrix}$.

The controllability matrix for the model is computed as: $M_C = [b, Ab] = \begin{bmatrix} 0 & 1 \\ 1 & -1 \end{bmatrix}$.

The pole placement controller gains are computed as: $\mathbf{k}^T = [0\ 3]$.
The resulting state feedback control law is given as: $f = -3v$.
The MATLAB commands for this example are given below:

```
A=[0 1;-10 -1]; B=[0;1];        % define state variable model
Mc=ctrb(A,B);                   % define controllability matrix
I=eye(size(A));                 % define identity matrix
cp=@(x)x^2+150*x+5000*eye(size(x)) % define anonymous function
K= I(end,:)/Mc*cp(A)            % contoller gains
```

We may note that the MATLAB codes for Examples 8.4 and 8.5 differ in the implementation of the matrix function. In Example 8.4, MALTB "polyvalm" function was used to define the matrix function; in Example 8.5, an anonymous function was defined to do the job.

## 8.1.5 Pole Placement using Sylvester's Equation

The Sylvester equation in algebra is a linear matrix equation, given as: $AX + XB = C$, where $A$ and $B$ are square matrices of not necessarily equal dimensions, and $C$ is a matrix of appropriate dimensions. The solution matrix $X$ is of the same diemension as $C$.

In order to apply the Sylvester equation for pole placement design, let an $n$-dimensional state variable model be given as:

$$\dot{\mathbf{x}}(t) = \mathbf{A}\mathbf{x}(t) + \mathbf{b}u(t), \quad y(t) = \mathbf{c}^T\mathbf{x}(t)$$

We assume that the pair $(\mathbf{A}, \mathbf{b})$ is controllable. Let $\mathbf{A} - \mathbf{b}\mathbf{k}^T$ represent the closed-loop system matrix; then, we perform a similarity transform on $\mathbf{A} - \mathbf{b}\mathbf{k}^T$ to express it as:

$$\mathbf{X}^{-1}(\mathbf{A} - \mathbf{b}\mathbf{k}^T)\mathbf{X} = \mathbf{\Lambda}$$

The matrix $\mathbf{\Lambda}$ on the right-hand-side is a diagonal matrix of desired eigenvalues, which can be alternatively assumed in modal form in the case of complex eigenvalues. Then, the Sylvester equation is written as:

$$\mathbf{A}\mathbf{X} - \mathbf{X}\mathbf{\Lambda} = \mathbf{b}\mathbf{k}^T\mathbf{X}$$

Next, let $\mathbf{k}^T\mathbf{X} = \mathbf{g}^T$ where $\mathbf{g}^T$ is any $1 \times n$ matrix so that $\mathbf{b}\mathbf{g}^T$ is $n \times n$. The resulting equation $\mathbf{A}\mathbf{X} - \mathbf{X}\mathbf{\Lambda} = \mathbf{b}\mathbf{g}^T$ can be solved for $\mathbf{X}$. A unique solution always exists as long as the eigenvalues of $\mathbf{A}$ and $\mathbf{\Lambda}$ are distinct, that is, $\lambda_i(\mathbf{A}) - \lambda_j(\mathbf{\Lambda}) \neq 0$.

The design procedure for pole placement using Sylvester's equation is given as follows:

1. Choose a matrix $\mathbf{\Lambda}$ of desired eigenvalues in modal form.
2. Choose any $\mathbf{g}^T$ and solve $\mathbf{A}\mathbf{X} - \mathbf{X}\mathbf{\Lambda} = \mathbf{b}\mathbf{g}^T$ for $\mathbf{X}$.
3. Recover the feedback gain vector $\mathbf{k}^T = \mathbf{g}^T\mathbf{X}^{-1}$.
4. If the solution $\mathbf{X}$ to the Sylvester equation is not invertible, choose a different $\mathbf{g}^T$ and repeat the procedure.

The application of the Sylvester equation for pole placement is illustrated using the following example.

**Example 8.6:** The mass–spring–damper model.

The state equation for the mass–spring–damper model is described as:

$$\frac{d}{dt}\begin{bmatrix} x \\ v \end{bmatrix} = \begin{bmatrix} 0 & 1 \\ -10 & -1 \end{bmatrix}\begin{bmatrix} x \\ v \end{bmatrix} + \begin{bmatrix} 0 \\ 1 \end{bmatrix}f$$

Let the desired characteristic polynomial be selected as: $\Delta_{des}(s) = s^2 + 4s + 10$; then, the modal matrix of desired eigenvalues is: $\mathbf{\Lambda} = \begin{bmatrix} -2 & \sqrt{6} \\ -\sqrt{6} & -2 \end{bmatrix}$.

Let $\mathbf{g}^T = [1 \ 0]$; then, $\mathbf{b}\mathbf{g}^T = \begin{bmatrix} 0 & 0 \\ 1 & 0 \end{bmatrix}$.

The resulting solution to Sylvester's equation is obtained as: $\mathbf{X} =$
$\begin{bmatrix} -0.067 & -0.082 \\ 0.333 & 0.0 \end{bmatrix}$.

The feedback controller gains are computed as: $\mathbf{k}^T = [0\ 3]$.

The MATLAB commands for this example are given below:

```
A=[0 1;-10 -1]; B=[0;1];        % define state variable model
Lam=[-2 sqrt(6);-sqrt(6) -2];   % define modal matrix
G=[1 0];                        % define arbitrary vector
X=sylvester(A,-Lam,B*g);        % solve sylvester equation
K=g/X                           % compute feedback gains
```

**Pole Placement Design in MATLAB.** The MATLAB Control System Toolbox includes the "place" command that uses the Ackermann's formula for pole placement design. The command is invoked by entering the system and input matrices, and a vector of desired roots of the characteristic polynomial.

The function returns the feedback gain vector that places the eigenvalues of the closed-loop system matrix, $\mathbf{A} - \mathbf{b}\mathbf{k}^T$, at the desired root locations.

Assuming that the state variable model $(\mathbf{A}, \mathbf{b})$ has been defined, let $p$ denote a vector of the desired closed-loop pole locations; then, the MATLAB "place" command is invoked as:

```
K=place(A, B, p)   % pole placement controller gains
```

In the case of DC motor model (Example 8.4), the returned controller gains are given as: $\mathbf{k}^T = \begin{bmatrix} 0.4 & 7.15 \end{bmatrix}$.

## 8.2 Tracking System Design

A tracking system is designed to reach and maintain zero steady-state error with respect to a reference input, $r(t)$. For a constant reference input, the robust design of a tracking system requires placing an integrator in the feedback loop.

Alternatively, a scaled version of constant reference input may be used to obtain zero steady-state error. This method, described first is, however, not robust to changes in plant parameters and should be used with discretion.

### 8.2.1 Tracking System Design with Feedforward Gain

In order to consider the tracking system design, let the state variable model be given as:

$$\dot{\mathbf{x}}(t) = \mathbf{A}\mathbf{x}(t) + \mathbf{b}u(t), \quad y(t) = \mathbf{c}^T\mathbf{x}(t)$$

The control law for tracking system design is defined as:

$$u = -\mathbf{k}^T x + k_r r,$$

where $k_r$ is a feedforward gain multiplying the reference input, $r(t)$. The resulting closed-loop system is given as:

$$\dot{\mathbf{x}}(t) = (\mathbf{A} - \mathbf{b}\mathbf{k}^T)\mathbf{x}(t) + \mathbf{b}k_r r(t), \quad y(t) = \mathbf{c}^T \mathbf{x}(t)$$

where $\dot{\mathbf{x}}(t) = \mathbf{0}$ denotes the steady-state of the system. Assuming closed-loop stability, the steady-state values of the state variables are obtained as:

$$\mathbf{x}_{ss} = -(\mathbf{A} - \mathbf{b}\mathbf{k}^T)^{-1}\mathbf{b}k_r r_{ss}.$$

The steady-state system output is given as:

$$y_{ss} = -\mathbf{c}^T (\mathbf{A} - \mathbf{b}\mathbf{k}^T)^{-1}\mathbf{b}\, k_r r_{ss}$$

Let $T(s) = \mathbf{c}^T(s\mathbf{I} - \mathbf{A} + \mathbf{b}\mathbf{k}^T)^{-1}\mathbf{b}$ represent the closed-loop transfer function; then, $y_{ss} = T(0)k_r r_{ss}$.

In order to ensure $y_{ss} = r_{ss}$, we may choose, $k_r = T(0)^{-1}$. Further, if $y = x_1$, then $k_r$ may be selected as $k_r = k_1$, to ensure $r - y = 0$ in the steady-state.

Tracking system design by feeding forward the reference input is illustrated in the following example.

**Example 8.7:** The DC motor model.

A pole placement controller for the DC motor model was designed as (Example 8.1) : $V_a = -0.4i_a - 7.15\omega$. We modify the control law to include a feedforward gain as: $V_a = -0.4i_a - 7.15\omega + k_r r$. Then, the closed-loop system model is given as:

$$\frac{d}{dt}\begin{bmatrix} i_a \\ \omega \end{bmatrix} = \begin{bmatrix} -140 & -720 \\ 5 & -10 \end{bmatrix}\begin{bmatrix} i_a \\ \omega \end{bmatrix} + \begin{bmatrix} 100 \\ 0 \end{bmatrix}k_r r$$

The state variables assume the following values in the steady-state:

$$\begin{bmatrix} i_{a,ss} \\ \omega_{ss} \end{bmatrix} = \begin{bmatrix} 0.2 \\ 0.1 \end{bmatrix}k_r r_{ss}.$$

The motor output is defined as: $\omega = \begin{bmatrix} 0 & 1 \end{bmatrix}\begin{bmatrix} i_a \\ \omega \end{bmatrix}$; then, the steady-state value of the output is given as: $y_{ss} = 0.1k_r r_{ss}$. Hence, we may choose

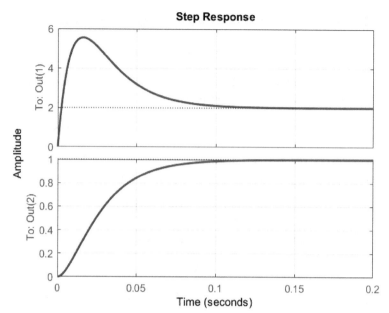

**Figure 8.2**  The step respone of the DC motor with pole placement controller and feedforward compensation: armature current (top); motor speed (bottom).

$k_r = 10$ for error-free tracking. The resulting control law is given as: $V_a = -0.4i_a - 7.15\omega + 10r$.

A plot of the motor response with feedforward compensation is shown in Figure 8.2. As expected, the motor speed settles at a value of unity in the steady-state. The MATLAB commands for computing the feedforward compensation are given as:

```
A=[-100 -5;5 -10]; B=[100;0]; % define state variable model
Gss=ss(A,B,eye(2),[0;0]);    % system model for state feedback
K=place(A, B, [-50, -100]);  % compute controller gains
Tss=[0 1]*feedback(Gss,K);   % closed-loop system
kr=1/dcgain(Tss);            % feedforward gain
step(A-B*K, B*kr,eye(2),[0;0]), grid % plot step response
```

**Example 8.8:** The mass–spring–damper model.

A pole placement controller for the mass–spring–damper model was designed as (Example 8.2): $f = -3v$. We modify the control law to include a feedforward gain as: $f = -3v + k_r r$. Then, the closed-loop system model is given as:

$$\frac{d}{dt}\begin{bmatrix} x \\ v \end{bmatrix} = \begin{bmatrix} 0 & 1 \\ -10 & -4 \end{bmatrix}\begin{bmatrix} x \\ v \end{bmatrix} + k_r r.$$

Assuming that the output equation is given as: $x = \begin{bmatrix} 1 & 0 \end{bmatrix} \begin{bmatrix} x \\ v \end{bmatrix}$, the steady-state output is: $y_{ss} = 0.1 k_r r_{ss}$. Hence, for error-free tracking, we may choose $k_r = 10$. The resulting control law is given as: $f = -3v + 10 r$.

The MATLAB commands for computing the feedforward compensation in this case are given as:

```
A=[0 1;-10 -1]; B=[0;1]; C=[1 0]; % define state variable model
Gss=ss(A,B,eye(2),[0;0]);         % system model for state feedback
K=place(A,B,-2+sqrt(6)*[j -j]);   % compute controller gains
Tss=C*feedback(Gss,K);            % closed-loop system
kr=1/dcgain(Tss)                  % feedforward gain
```

Next, we consider the design of reference tracking controllers for state variable models by integrating the error signal.

## 8.2.2 Tracking PI Controller Design

The tracking system design by using a feedforward gain to cancel the tracking error is not robust to changes in plant parameters. In a more general context, if elimination of the steady-state error is desired, then an integral controller should be added to the feedback loop (Figure 8.3). The integral controller integrates the error signal, thereby forcing it to zero in the steady-state.

The proportional-integral (PI) type control law for tracking system design using state feedback is defined as:

$$u = -\mathbf{k}^T \mathbf{x} + k_i \int (r - y)\, dt$$

The input to the integrator is the error signal: $e = r - y$, where $r(t)$ represents a time-varying reference input. The integrator output denotes an additional state variable, $x_a(t)$; then, the integrator state equation is given as:

$$\dot{x}_a = r - y = r - \mathbf{c}^T \mathbf{x}.$$

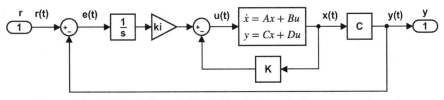

**Figure 8.3** The block diagram for tracking PI control of a state variable model.

By adding the integrator output to the state variables, the augmented system model has $n + 1$ variables and is described as:

$$\begin{bmatrix} \dot{\mathbf{x}} \\ \dot{x}_a \end{bmatrix} = \begin{bmatrix} \mathbf{A} & 0 \\ -\mathbf{c}^T & 0 \end{bmatrix} \begin{bmatrix} \mathbf{x} \\ x_a \end{bmatrix} + \begin{bmatrix} \mathbf{b} \\ 0 \end{bmatrix} u + \begin{bmatrix} 0 \\ 1 \end{bmatrix} r.$$

A full-state feedback control law for the augmented system is defined as:

$$u = \begin{bmatrix} -\mathbf{k}^T & k_i \end{bmatrix} \begin{bmatrix} \mathbf{x} \\ x_a \end{bmatrix},$$

where $k_i$ represents the integral gain. After substituting the above control law in the augmented system model, the closed-loop system is described as:

$$\begin{bmatrix} \dot{\mathbf{x}} \\ \dot{x}_a \end{bmatrix} = \begin{bmatrix} \mathbf{A} - \mathbf{b}\mathbf{k}^T & bk_i \\ -\mathbf{c}^T & 0 \end{bmatrix} \begin{bmatrix} \mathbf{x} \\ x_a \end{bmatrix} + \begin{bmatrix} 0 \\ 1 \end{bmatrix} r$$

The characteristic polynomial of the above system is computed as:

$$\Delta_a(s) = \begin{vmatrix} s\mathbf{I} - \mathbf{A} + \mathbf{b}\mathbf{k}^T & -bk_i \\ \mathbf{c}^T & s \end{vmatrix}, \text{ where } \mathbf{I} \text{ denotes an identity matrix of}$$

order $n$.

Next, we choose a desired $(n + 1)$ order characteristic polynomial, $\Delta_{des}(s)$, and perform the pole placement design for the augmented system. The location of the integrator pole may be selected by trial and error keeping in view the desired settling time of the system.

The design of the tracking PI controller is illustrated in the following examples.

**Example 8.9:** The mass–spring–damper system.

Consider the mass–spring–damper model described by the following state variable model:

$$\frac{d}{dt} \begin{bmatrix} x \\ v \end{bmatrix} = \begin{bmatrix} 0 & 1 \\ -10 & -1 \end{bmatrix} \begin{bmatrix} x \\ v \end{bmatrix} + \begin{bmatrix} 0 \\ 1 \end{bmatrix} f, x = \begin{bmatrix} 1 & 0 \end{bmatrix} \begin{bmatrix} x \\ v \end{bmatrix}.$$

where it is desired to have less than $10\%$ overshoot and zero steady-state error to a step input.

In order to perform integral control, an augmented state variable model is formed as:

$$\frac{d}{dt} \begin{bmatrix} x \\ v \\ x_a \end{bmatrix} = \begin{bmatrix} 0 & 1 & 0 \\ -10 & -1 & 0 \\ -1 & 0 & 0 \end{bmatrix} \begin{bmatrix} x \\ v \\ x_a \end{bmatrix} + \begin{bmatrix} 0 \\ 1 \\ 0 \end{bmatrix} u + \begin{bmatrix} 0 \\ 0 \\ 1 \end{bmatrix} r.$$

The control law for the augmented system is given as: $u = -k_1 x - k_2 v + k_i \int (r - x) \, dt$.

The resulting closed-loop characteristic polynomial is obtained as: $\Delta(s) = s^3 + (k_2 + 1)s^2 + (k_1 + 10)s + k_i$.

Let a third-order desired characteristic polynomial be selected as: $\Delta_{des}(s) = (s + 1)(s^2 + 4s + 10)$.

For comparison, two other choices of characteristic polynomials, that is, $\Delta_{des}(s) = (s + 2)(s^2 + 4s + 10)$ and $\Delta_{des}(s) = (s + 3)(s^2 + 4s + 10)$, are also considered. The corresponding controller gains are tabulated below. The settling times of the unit-step responses of the closed-loop systems are also reported.

A comparison of the settling times shows that an optimal gain to achieve minimum settling time for the integral controller lies between $k_i = 20$ and $k_i = 30$.

| $\Delta_{des}(s)$ | Controller Gains | Setting Time |
|---|---|---|
| $(s + 1)(s^2 + 4s + 10)$ | $k_1 = 4, \ k_2 = 4, \ k_i = 10.$ | $4.27s$ |
| $(s + 2)(s^2 + 4s + 10)$ | $k_1 = 8, \ k_2 = 5, \ k_i = 20.$ | $1.9s$ |
| $(s + 3)(s^2 + 4s + 10)$ | $k_1 = 12, \ k_2 = 6, \ k_i = 30.$ | $2.1s$ |

The MATLAB commands for computing the controller gains and plotting the step response for the mass–spring–damper system are given below (Figure 8.4). Design of a sample controller is shown; the other designs are similarly performed.

```
A=[0 1; -10 -1]; B=[0;1]; C=[1 0]; % define state variable model
Aa=[A [0;0];-C 0]; Ba=[B;0];       % define augmented system model
K=place(Aa,Ba,[-1 -2+sqrt(6)*[j -j]]); % pole placement controller
T=ss(Aa-Ba*K1,[0;B],[C 0],0);      % obtain the closed-loop system
step(T),grid                       % plot step response
```

**Example 8.10:** The DC motor model

The state variable model for a small DC motor is given as:

$$\frac{d}{dt} \begin{bmatrix} i_a \\ \omega \end{bmatrix} = \begin{bmatrix} -100 & -5 \\ 5 & -10 \end{bmatrix} \begin{bmatrix} i_a \\ \omega \end{bmatrix} + \begin{bmatrix} 100 \\ 0 \end{bmatrix} V_a$$

$$\omega = \begin{bmatrix} 0 & 1 \end{bmatrix} \begin{bmatrix} i_a \\ \omega \end{bmatrix}$$

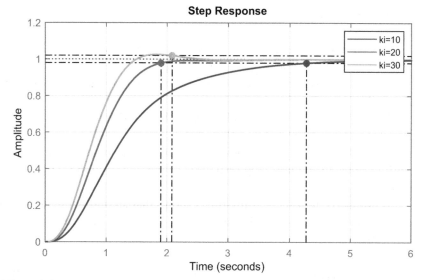

**Figure 8.4** Step response of the mass–spring–damper system with integrator in the loop.

The control law for the tracking PI controller for DC motor is given as:
$u = -k_1 i_a - k_2 \omega + k_i \int (r - \omega) dt$.

The augmented system model for pole placement design using integral control is obtained as:

$$\frac{d}{dt}\begin{bmatrix} i_a \\ \omega \\ x_a \end{bmatrix} = \begin{bmatrix} -100 & -5 & 0 \\ 5 & -10 & 0 \\ 0 & -1 & 0 \end{bmatrix}\begin{bmatrix} i_a \\ \omega \\ x_a \end{bmatrix} + \begin{bmatrix} 100 \\ 0 \\ 0 \end{bmatrix} u + \begin{bmatrix} 0 \\ 0 \\ 1 \end{bmatrix} r.$$

The resulting closed-loop characteristic polynomial is given as:

$$\Delta(s) = s^3 + (100k_1 + 100)s^2 + (1000k_1 + 500k_2 + 1025)s - 500k_i.$$

We may choose a desired characteristic polynomial $\Delta_{des}(s)$ with closed-loop roots located, e.g., at: $s = -50, -75, -100$, which results in the controller gains: $k_1 = 1.9, k_2 = 49.15, k_i = -1500$.

For comparison, we consider alternate choices of characteristic polynomials with complex roots. In the first case, roots are located along a vertical line at: $s = -70.7, -70.7 \pm j70.7$. In the second case, the roots are located along a circle at: $s = -70.7, -65.3 \pm j27$.

The resulting feedback gains and the settling times for the closed-loop system step responses are displayed in the following table. The step responses of the closed-loop systems are compared below (Figure 8.5).

| $\Delta_{des}(s)$ | Controller Gains | Setting Time |
|---|---|---|
| $(s+50)(s+75)(s+100)$ | $k_1 = 1.9, k_2 = 49.15,\ k_i = -1500$ | 112 ms |
| $(s+70.7)(s^2+141.4s+10,000)$ | $k_1 = 1.02, k_2 = 35.9,\ k_i = -1414$ | 96 ms |
| $(s+70.7)(s^2+130.7s+5,000)$ | $k_1 = 1.13, k_2 = 27.2,\ k_i = -707$ | 61 ms |
| $(s+9.94)(s^2+100s+5,030)$ | $K(s) = \frac{10(s+10)}{s}$ | 82 ms |

Further, the transfer function model of the DC motor model is obtained as: $G(s) = \frac{500}{s^2+110s+1025}$. Let a PI controller for the model be defined as: $K(s) = \frac{K(s+10)}{s}$.

Then, using the root locus technique, we select a controller gain: $K = 10$, which results in closed-loop roots at: $s = -50 \pm j50.4, -9.94$. The resulting closed-loop system step response has a settling time of 82 ms, which is comparable with full-state feedback controllers designed above. The step response of the PI controller is also plotted below (Figure 8.5).

The MATLAB script for the DC motor example is given below. State space design of a sample controller is presented; the other designs are similarly performed.

```
% state space design
A=[-100 -5;5 -10]; B=[100;0]; C=[0 1]; % define system model
Aa=[A [0;0];-C 0]; Ba=[B;0];     % define augmented system model
K=place(Aa,Ba, -100/sqrt(2)*exp([0 j*pi/8 -j*pi/8])); % controlr
T=ss(Aa-Ba*K1,[0;0;1],[C 0],0);  % obtain closed-loop system
step(T),grid                      % plot step response
% transfer function design
```

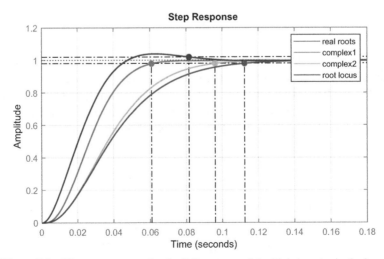

**Figure 8.5** The step response for the DC motor model with integrator in the loop.

```
G=tf(500,[1 110 1025]);        % define TF model
 Kpi=tf([1 10],[1 0]);          % define PI controller
 T=feedback(10*Kpi*G,1);        % closed-loop system
 step(T),grid                   % plot step response
```

## 8.3 State Variable Models of Sampled-Data Systems

System models described in state variable form can be converted to their discrete-time equivalents by considering the effect of a ZOH device at the input of the model. The ZOH device models a digital-to-analog converter that converts the number sequence into a continuous-time signal by holding its output constant for successive time periods.

To develop this approach, let the continuous-time state variable model be given as:

$$\dot{\mathbf{x}}(t) = \mathbf{A}\mathbf{x}(t) + \mathbf{B}u(t)$$
$$y(t) = \mathbf{C}\mathbf{x}(t).$$

where $\mathbf{A}$ is the system matrix, $\mathbf{B}$ is the input matrix, and $\mathbf{C}$ is the output matrix.

### 8.3.1 Discretizing the State Equations

In order to discretize the state equations, we recal that the time-domain solution to the state equation (Chapter 3) was given as:

$$\mathbf{x}(t) = e^{\mathbf{A}(t-t_0)}\mathbf{x}_0 + \int_{t_0}^{t} e^{\mathbf{A}(t-\tau)}\mathbf{B}u(\tau)\,d\tau.$$

We assume that the input to the plant, generated by the ZOH, is held constant over successive time periods, $(k-1)T \le t < kT, k = 1, 2, \ldots$.

We further assume that the system state is available at $t_0 = (k-1)T$ and wish to construct it at $t = kT$ using the above convolution integral; using subscript to denote discretized time, the result is given as:

$$\mathbf{x}_k = e^{\mathbf{A}t}\mathbf{x}_{k-1} + \int_{(k-1)T}^{kT} e^{\mathbf{A}T}\mathbf{B}u_k\,d\tau$$

A simple change of variables results in the following expression:

$$\mathbf{x}_k = e^{\mathbf{A}t}\mathbf{x}_{k-1} + \int_{0}^{T} e^{\mathbf{A}T}\,d\tau\mathbf{B}\,u_k$$

Next, we define the following system and input matrices for the discrete-time model:

$$\mathbf{A}_d = e^{\mathbf{A}T}, \quad \mathbf{B}_d = \int_0^T e^{\mathbf{A}\tau} d\tau \mathbf{B}$$

Then, given a continuous-time state variable model, its corresponding discrete-time model is obtained as:

$$\mathbf{x}_{k+1} = \mathbf{A}_d \mathbf{x}_k + \mathbf{B}_d u_k, \quad y_k = \mathbf{C}_d \mathbf{x}_k.$$

where $\mathbf{C}_d = \mathbf{C}$. We note that the matrices $(\mathbf{A}_d, \mathbf{B}_d)$ appearing in the discrete-time model are parameterized by $T$. Further, assuming that system matrix $A$ is invertible, the expression for $\mathbf{B}_d$ can be simplified as:

$$\mathbf{B}_d = \left[ \int_0^T \left( I + \mathbf{A}\tau + \frac{\mathbf{A}^2\tau^2}{2!} + \cdots \right) d\tau \right]$$

$$= \left( IT + \frac{\mathbf{A}T^2}{2!} + \cdots \right) \mathbf{B} = \mathbf{A}^{-1}(e^{\mathbf{A}T} - \mathbf{I})\mathbf{B}.$$

**Discretization of State Variable Models in MATLAB.** In the MATLAB Control Systems Toolbox, the "c2d" is used for the discretization of a given state variable model, where the presence of a ZOH at the input is assumed. The command is invoked as:

```
[Ad,Bd]=c2d(A,B,Ts)          % obtain DT state variable model
```

The 'c2d' command permits using alternate discretization methods (these were explored in Example 7.24).

The discretization of a state variable model is illustrated via the following example.

**Example 8.11:** The DC motor model.

The state-space model of a small DC motor is given as:

$$\frac{d}{dt} \begin{bmatrix} i_a \\ \omega \end{bmatrix} = \begin{bmatrix} -100 & -5 \\ 5 & -10 \end{bmatrix} \begin{bmatrix} i_a \\ \omega \end{bmatrix} + \begin{bmatrix} 100 \\ 0 \end{bmatrix} V_a$$

$$\omega = \begin{bmatrix} 0 & 1 \end{bmatrix} \begin{bmatrix} i_a \\ \omega \end{bmatrix}.$$

Let $T = 0.02s$, selected in view of the dominant motor time constant: $\tau_m \cong 0.1s$.; then, the system and input matrices for the discrete model are computed as:

$$\mathbf{A}_d = e^{\mathbf{A}t} = \begin{bmatrix} 0.134 & -0.038 \\ 0.038 & 0.816 \end{bmatrix}, \quad \mathbf{B}_d = \mathbf{A}^{-1}(e^{\mathbf{A}t} - \mathbf{I})\mathbf{B} = \begin{bmatrix} 0.863 \\ 0.053 \end{bmatrix}.$$

The resulting discrete state variable model is given as:

$$\begin{bmatrix} i_{k+1} \\ \omega_{k+1} \end{bmatrix} = \begin{bmatrix} 0.134 & -0.038 \\ 0.038 & 0.816 \end{bmatrix} \begin{bmatrix} i_k \\ \omega_k \end{bmatrix} + \begin{bmatrix} 0.863 \\ 0.053 \end{bmatrix} V_k$$

$$y_k = \begin{bmatrix} 0 & 1 \end{bmatrix} \begin{bmatrix} i_k \\ \omega_k \end{bmatrix}.$$

We may note that the discretized $A_d$ matrix has eigenvalues located at: $z_{1,2} = 0.814, 0.136$, which are related to the analog system eigenvalues: $s_{1,2} = -99.7, -10.28$ by the relation: $z = e^{Ts}$.

In the MATLAB Control Systems Toolbox, the state variable model of the DC motor is discretized by using the "c2d" command as follows:

```
A=[-100 -5; 5 -10]; B=[100; 0]; % define CT state variable model
[Ad,Bd]=c2d(A,B,.02)           % obtain DT state variable model
```

Alternately, starting with the transfer function model of the DC motor, the "ss" command in the MATLAB Control Systems Toolbox may be used to realize it into a state variable model in the controller form, which is then discretized to obtain an equivalent discrete-time state variable model of the DC motor. This is done by issuing the following commands:

```
G=tf(500,[1 110 1025]);        % define DC motor transfer function
Gz=c2d(ss(G),.02)              % obtain DT state variable model
```

The resulting discrete-time model is given as:

$$\begin{bmatrix} x_{1,k+1} \\ x_{2,k+1} \end{bmatrix} = \begin{bmatrix} 0.058 & -0.243 \\ 0.243 & 0.892 \end{bmatrix} \begin{bmatrix} x_{2,k} \\ x_{2,k} \end{bmatrix} + \begin{bmatrix} 0.030 \\ 0.013 \end{bmatrix} u_k$$

$$y_k = \begin{bmatrix} 0 & 3.91 \end{bmatrix} \begin{bmatrix} x_{1,k} \\ x_{2,k} \end{bmatrix}.$$

We note that the state variables of the two models are different, resulting in different structures for the models. Both models, however, share the same $z$-plane eigenvalues: $z_{1,2} = 0.814, 0.136$, as well as the pulse transfer function, given as:

$$G(z) = \frac{0.0526(z + 0.488)}{(z - 0.136)(z - 0.814)}.$$

## 8.3.2  Solution to the Discrete State Equations

The discrete-time state equations constitute a set of first-order difference equations that can be easily solved by iteration by assuming an initial state vector and an input sequence.

Toward this end, let the discrete-time state equation be given as:

$$\mathbf{x}_{k+1} = \mathbf{A}_d\mathbf{x}_k + \mathbf{B}_d u_k, \quad y_k = \mathbf{C}_d\mathbf{x}_k.$$

Then, given an initial vector, $x_0$, and an input sequence $u\{k\}$, an iterative solution to the discrete state equation is developed as follows:

$$\mathbf{x}_1 = \mathbf{A}_d\mathbf{x}_0 + \mathbf{B}_d u_0$$
$$\mathbf{x}_2 = \mathbf{A}_d^2\mathbf{x}_0 + \mathbf{A}_d\mathbf{B}_d u_0 + \mathbf{B}_d u_1$$
$$\vdots$$
$$\mathbf{x}_n = \mathbf{A}_d^n\mathbf{x}_0 + \sum_{k=0}^{n-1} \mathbf{A}_d^{n-1-k}\mathbf{B}_d u_k.$$

The state transition matrix in the discrete-time case is defined as: $\Phi(k) = \mathbf{A}_d^k$. In terms of the state transition matrix, the state vector evolves as:

$$\mathbf{x}_n = \Phi(n)\mathbf{x}_0 + \sum_{k=0}^{n-1} \Phi(n-1-k)\mathbf{B}_d u_k.$$

The iterative solution to the discrete state equations is illustrated by the following example:

**Example 8.12:** The DC motor model.

The discrete state variable model of a small DC motor ($T = 0.02s$) is described by:

$$\mathbf{A}_d = \begin{bmatrix} 0.134 & -0.038 \\ 0.038 & 0.816 \end{bmatrix}, \quad \mathbf{B}_d = \begin{bmatrix} 0.863 \\ 0.053 \end{bmatrix}, \quad \mathbf{C}_d = \begin{bmatrix} 0 & 1 \end{bmatrix}$$

We assume that the initial conditions are zero, and the input to the model is a unit-step sequence: $u_k = \{1, 1, \ldots\}$; then, the output sequence is iteratively computed in MATLAB as follows:

```
A=[-100 -50; 5 -10]; B=[100; 0]; C=[0 1]; % define dc motor model
[Ad,Bd]=c2d(A,B,.02);          % obtain DT state variable model
x=zeros(2,10);                 % initialize the solution matrix
for k=2:10, x(:,k)= Ad*x(:,k-1)+Bd; end % iterative solution
y=C*x;                         % obtain output sequence
```

The results of the above iteration are given as:

$$y\{k\} = \{0, 0.053, 0.128, 0.194, 0.249, 0.293, 0.329, 0.359, 0.383,$$
$$0.402, 0.418, \ldots\}$$

### 8.3.3 Pulse Transfer Function from State Equations

The pulse transfer function, $G(z)$, of the sampled-data system can be obtained from the discrete state variable description by the application of $z$-transform, which results in:

$$z\mathbf{x}(z) - z\mathbf{x}_0 = \mathbf{A}_d\mathbf{x}(z) + \mathbf{B}_d u(z)$$

The above equation is solved for the state vector assuming zero initial conditions to obtain:

$$\mathbf{x}(z) = (z\mathbf{I} - \mathbf{A}_d)^{-1}\mathbf{B}_d u(z)$$

The output equation is given as:

$$y(z) = \mathbf{C}_d(z\mathbf{I} - \mathbf{A}_d)^{-1}\mathbf{B}_d u(z) = G(z)u(z).$$

Hence, given the discrete state equations, the pulse transfer function is obtained as:

$$G(z) = \mathbf{C}_d(z\mathbf{I} - \mathbf{A}_d)^{-1}\mathbf{B}_d.$$

The state transition matrix for the sampled-data system is obtained by taking the inverse $z$-transform of $(z\mathbf{I} - \mathbf{A}_d)^{-1}$ and is given as:

$$\phi(k) = z^{-1}\{(z\mathbf{I} - \mathbf{A}_d)^{-1}\} = \mathbf{A}_d^k.$$

In terms of the state transition matrix, the unit pulse response the sampled-data system is obtained as:

$$g_k = \mathbf{C}_d\mathbf{A}_d^{k-1}\mathbf{B}, \quad k \geq 0.$$

**Pulse Transfer Function in MATLAB.** The pulse transfer function for a given discrete state variable model can be obtained in the MATLAB Control System Toolbox by invoking the "tf" command. The MATLAB commands used for this purpose are given below:

```
Gz=ss(A,B,C,0,T);        % define DT state variable model
tf(Gz)                   % obtain transfer function
```

The following example illustrates the use of the above MATLAB commands.

**Example 8.13:** The DC motor model.

The discrete-time state variable model of a small DC motor ($T = 0.02s$) is given as:

$$\mathbf{A}_d = \begin{bmatrix} 0.134 & -0.038 \\ 0.038 & 0.816 \end{bmatrix}, \quad \mathbf{B}_d = \begin{bmatrix} 0.863 \\ 0.053 \end{bmatrix}, \quad \mathbf{C}_d = \begin{bmatrix} 0 & 1 \end{bmatrix}.$$

Then, by using the above commands, the motor transfer function is obtained as:

$$G(z) = \frac{0.053z + 0.0257}{z^2 - 0.95z + 0.111}.$$

## 8.4 Controllers for Discrete State Variable Models

In this section, we discuss the design of digital controllers for the state variable models of sampled-data systems. We begin with the emulation of analog pole placement controller for the sampled-data system model, followed by direct design of pole placement controller for a discrete-time system model.

### 8.4.1 Emulating an Analog Controller

The analog pole placement controller designed for the continuous-time state variable model can be incorporated into the equivalent sampled-data system model. Successful controller emulation requires a high enough sampling rate that is at least ten times the frequency of the dominant closed-loop system poles.

We consider the emulation of pole placement controller designed for the DC motor model (Example 8.1) for controlling the discrete-time model of the DC motor. The DC motor model is discretized at two different sampling rates for comparison, assuming ZOH at the plant input.

**Example 8.14:** The DC motor model.

The state and output equations for a small DC motor model are given as:

$$\frac{d}{dt}\begin{bmatrix} i_a \\ \omega \end{bmatrix} = \begin{bmatrix} -100 & -5 \\ 5 & -10 \end{bmatrix} \begin{bmatrix} i_a \\ \omega \end{bmatrix} + \begin{bmatrix} 100 \\ 0 \end{bmatrix} V_a, \quad \omega = \begin{bmatrix} 0 & 1 \end{bmatrix} \begin{bmatrix} i_a \\ \omega \end{bmatrix}.$$

In Example 8.1, the state feedback controller gains for the desired characteristic polynomial: $\Delta_{des}(s) = s^2 + 150s + 5000$, were computed as: $\mathbf{k}^T = \begin{bmatrix} 0.4 & 7.15 \end{bmatrix}$. We use the same controller gains to control the corresponding sample-data system model of the DC motor. The motor model is discretized at two different sampling rates for this purpose. The results are:

$$T = 0.01s : \mathbf{A}_d = \begin{bmatrix} 0.367 & -0.030 \\ 0.030 & 0.904 \end{bmatrix}, \quad \mathbf{B}_d = \begin{bmatrix} 0.632 \\ 0.018 \end{bmatrix}, \quad \mathbf{C}_d = \begin{bmatrix} 0 & 1 \end{bmatrix}.$$

$$T = 0.02s : \mathbf{A}_d = \begin{bmatrix} 0.134 & -0.038 \\ 0.038 & 0.816 \end{bmatrix}, \quad \mathbf{B}_d = \begin{bmatrix} 0.863 \\ 0.053 \end{bmatrix}, \quad \mathbf{C}_d = \begin{bmatrix} 0 & 1 \end{bmatrix}.$$

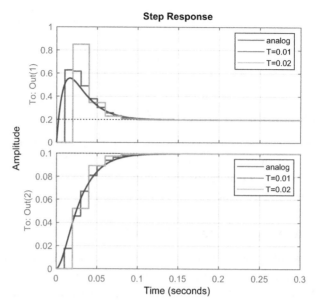

**Figure 8.6**    The step response of the DC motor with controller emulation: armature current (top); motor speed (bottom).

The resulting closed-loop system models are simulated with a unit-step input, and both state variables, $i_a(t)$ and $\omega(t)$, are plotted (Figure 8.6). The response of the continuous-time system is plotted alongside for comparison.

The MATLAB script for this example is given below:

```
A=[-100 -50; 5 -10]; B=[100; 0]; C=[0 1]; % define dc motor model
K=place(A, B, [-50, -100]);   % feedback controller gains
Gss=ss(A,B,eye(2),[0;0]);     % dynamic system object
Tss=feedback(Gss,K);          % closed-loop system
Gd1=c2d(Gss,.01);             % discrete model 1
Gd2=c2d(Gss,.02);             % discrete model 2
Td1=feedback(Gd1,K);          % closed-loop system 1
Td2=feedback(Gd2,K);          % closed-loop system 2
step(Tss,Td1,Td2),grid        % plot step response
legend('analog','discrete1','discrete2') % figure legend
```

We observe from the figure that the overshoot in the armature current increases for the lower sampling rate, though both models display similar settling time of about 100 m sec.

### 8.4.2 Pole Placement Design of Digital Controller

Given a discrete-time system model in state variable form, $\{A_d, B_d\}$, and a desired pulse characteristic polynomial $\Delta_{des}(z)$, a state feedback controller for the system can be designed using pole placement similar to that of the continuous-time system (Section 8.1).

To proceed further, let the discrete-time SISO system model be given as:

$$\mathbf{x}_{k+1} = \mathbf{A}_d \mathbf{x}_k + b_d u_k$$
$$y_k = \mathbf{c}^T \mathbf{x}_k.$$

A digital state feedback controller for the discrete-time model is defined as:

$$u_k = -\mathbf{k}^T \mathbf{x}_k.$$

where $\mathbf{k}^T$ represents a row vector of constant feedback gains. These gains can be solved by equating the coefficients of the characteristic polynomial with those of a desired polynomial:

$$\Delta(z) = |z\mathbf{I} - \mathbf{A}_d| = \Delta_{des}(z).$$

The $\Delta_{des}(z)$ may be selected as a stable polynomial (in $z$), with roots located inside the unit circle that satisfy given damping ratio and/or settling time requirements. Assuming that the desired $s$-plane root locations are known, the corresponding $z$-plane root locations can be obtained from the equivalence: $z = e^{Ts}$.

The controller design process is illustrated in the following example.

**Example 8.15:** The DC motor model.

The discrete-time model of a small DC motor is given as ($T = 0.02s$):

$$\begin{bmatrix} i_{k+1} \\ \omega_{k+1} \end{bmatrix} = \begin{bmatrix} 0.134 & -0.038 \\ 0.038 & 0.816 \end{bmatrix} \begin{bmatrix} i_k \\ \omega_k \end{bmatrix} + \begin{bmatrix} 0.863 \\ 0.053 \end{bmatrix} V_k$$

$$y_k = \begin{bmatrix} 0 & 1 \end{bmatrix} \begin{bmatrix} i_k \\ \omega_k \end{bmatrix}.$$

The desired $s$-plane root locations for this model were earlier selected as: $s = -50, -100$. The corresponding $z$-plane roots ($T = 0.02s$) are obtained as: $z = e^{-1}, e^{-2}$.

The desired characteristic polynomial is given as: $\Delta_{des}(z) = z^2 - 0.95z + 0.05$.

The feedback gains $\mathbf{k}^T = [k_1, k_2]$, computed using the MATLAB "place" command, are given as: $k_1 = 0.247$, $k_2 = 4.435$.

An update rule for implementation of the controller on computer is obtained as:

$$u_k = 0.247 \, i_k + 4.435 \, \omega_k.$$

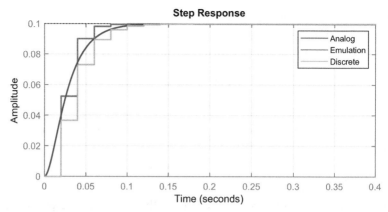

**Figure 8.7**   Unit-step response of the DC motor model: analog system; controller emulation; digital controller.

The step response of the closed-loop system is plotted below (Figure 8.7). The step response of the continuous-time system and that for the emulated controller gains (Example 8.14) are plotted alongside; the discrete system response was scaled to match the analog system response.

The MATLAB script for this example is given below:

```
A=[-100 -5; 5 -10]; B=[100;0]; C=[0 1]; % define DC motor model
[Ad,Bd]=c2d(A,B,.02);              % discretize DC motor model
K=place(A,B, [-100 -50]);          % pole placement for CT model
Kd=place(Ad,Bd, exp([-1 -2]));     % pole placement for DT model
Tss=ss(A-B*K,B,C,0);               % CT closed-loop system
Td1=ss(Ad-Bd*K,Bd,C,0,.02);        % DT closed-loop system 1
Td2=ss(Ad-Bd*Kd,Bd,C,0,.02);       % DT closed-loop system 2
gc= dcgain(Tss);                   % dc gain of continuous system
gd=dcgain(Td2);                    % dc gain of discrete system
step(Tss,Td1,gc/gd*Td2), grid      % plot step response
legend('Analog', 'Emulation', 'Discrete') % figure legend
```

### 8.4.3 Deadbeat Controller Design

A discrete-time system is called deadbeat if all closed-loop poles are placed at the origin $(z = 0)$. A deadbeat system has the remarkable property that its response reaches steady-state in $n$-steps, where $n$ represents the model dimension. This property makes it appealing to choose the desired characteristic polynomial as $\Delta_{des}(z) = z^n$, or equivalently, $T(z) = z^{-n}$.

To design a deadbeat controller using state feedback, let the closed-loop pulse transfer function be defined as:

$$T(z) = \frac{K(z)G(z)}{1 + K(z)G(z)}.$$

The above equation is solved for $K(z)$ to obtain:

$$K(z) = \frac{1}{G(z)} \frac{T(z)}{1 - T(z)}.$$

Let $T(z) = z^{-n}$; then, $K(z) = \frac{1}{G(z)(z^n - 1)}$.

The design of the deadbeat controller is illustrated in the following examples.

**Example 8.16:** Let $G(s) = \frac{1}{s+1}$; then $G(z) = \frac{1 - e^{-T}}{z - e^{-T}}$.

A deadbeat controller for the model is obtained as: $K(z) = \frac{z - e^{-T}}{(1 - e^{-T})(z - 1)}$.

**Example 8.17:** The DC motor model.

The discrete-time DC motor model, obtained via the MATLAB "c2d" command for $T = 0.02\ s$ is given as:

$$\begin{bmatrix} i_{k+1} \\ \omega_{k+1} \end{bmatrix} = \begin{bmatrix} 0.134 & -0.038 \\ 0.038 & 0.816 \end{bmatrix} \begin{bmatrix} i_k \\ \omega_k \end{bmatrix} + \begin{bmatrix} 0.863 \\ 0.053 \end{bmatrix} V_k$$

$$y_k = \begin{bmatrix} 0 & 1 \end{bmatrix} \begin{bmatrix} i_k \\ \omega_k \end{bmatrix}.$$

The desired characteristic polynomial is selected as: $\Delta_{des}(z) = z^2$.

The MATLAB "place" command cannot be used in the case due to the presence of repeated poles. However, by inserting $u_k = -\begin{bmatrix} k_1 & k_2 \end{bmatrix} \mathbf{x}_k$, direct calculations result in the following closed-loop characteristic polynomial:

$$\Delta(z) = z^2 + (0.863k_1 + 0.053k_2 - 0.95)z - 0.707k_1 + 0.026k_2 + 0.111$$

By equating the second and third term in the polynomial to zero, deadbeat controller gains are obtained as: $k_1 = 0.501$, $k_2 = 9.702$. The update rule for controller implementation is given as:

$$u_k = 0.501i_k + 9.702\omega_k.$$

The step response of the deadbeat controller shows no overshoot as the response settles in two time periods (Figure 8.8). The step response of the

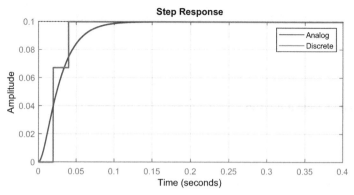

**Figure 8.8**   The step response of a small DC motor model: analog system; deadbeat controller.

continuous-time system (scaled to match the discrete system response) is shown alongside in the figure.

An approximate deadbeat design can be performed in the MATLAB Control Systems Toolbox by choosing distinct closed-loop eigenvalues close to the origin, such as: $z = \pm 10^{-5}$. The state feedback gains for the approximate design are obtained as: $k_1 = 0.509$, $k_2 = 9.702$, which are similar to the values calculated above. The resulting closed-loop system response is still deadbeat.

The MATLAB script for the approximate deadbeat design is given below; the closed-loop system response is plotted (in Figure 8.8), where the deadbeat response was adjusted to match the analog system response.

```
A=[-100 -5; 5 -10]; B=[100;0]; C=[0 1]; % define DC motor model
[Ad,Bd]=c2d(A,B,.02);              % discretize the motor model
K=place(A,B, [-100 -50]);          % pole placement for CT model
Kd=place(Ad,Bd,1e-5*[-1 1])        % pole placement for DT model
Tss=ss(A-B*K,B,C,0);               % CT closed-loop system
Td=ss(Ad-Bd*Kd,Bd,C,0,.02);        % DT closed-loop system
gc= dcgain(Tss);                   % dc gain of continuous system
gd=dcgain(Td);                     % dc gain of discrete system
step(Tss,gc/gd*Td),grid            % step response
legend('Analog','Discrete')        % figure legend
```

An exact deadbeat design can be performed using analytical computations with the help of MATLAB Symbolic Math Toolbox as shown in the following MATLAB script.

```
syms k1 k2 real                    % define symbolic variables
sympref('FloatingPointOutput',true); % define output format
Adcl=Ad-Bd*[k1 k2];                % define closed-loop system
pc=charpoly(Adcl);                 % characteristic polynomial
```

```
S=solve(pc(2),pc(3));          % solve for controller gains
Kd=double([S.k1, S.k2]);       % obtain numerical values
Td=ss(Ad-Bd*Kd,Bd,C,0,.02);    % DT closed-loop system
zpk(Td)                        % closed-loop transfer function
```

### 8.4.4 Tracking PI Controller Design

A tracking PI controller for the discretized state variable model can be designed similar to the continuous-time system design. Accordingly, we place an integrator in the feedback loop that integrates the error signal, thus ensuring that the tracking error goes to zero in the steady-state.

Toward this end, the discrete integrator output, obtained by backward difference approximation is given as: $v_k = v_{k-1} + T e_k$. The augmented system model including the integrator state variable is given as:

$$\begin{bmatrix} \mathbf{x}(k+1) \\ v(k+1) \end{bmatrix} = \begin{bmatrix} \mathbf{A}_d & 0 \\ -\mathbf{c}^T T & 1 \end{bmatrix} \begin{bmatrix} \mathbf{x}(k) \\ v(k) \end{bmatrix} + \begin{bmatrix} \mathbf{b}_d \\ 0 \end{bmatrix} u(k) + \begin{bmatrix} 0 \\ T \end{bmatrix} r(k).$$

The state feedback controller for the augmented system is given as:

$$u(k) = \begin{bmatrix} -\mathbf{k}^T & k_i \end{bmatrix} \begin{bmatrix} x(k) \\ v(k) \end{bmatrix},$$

where $k_i$ represents the integral gain. With the addition of the above controller, the closed-loop system is described as:

$$\begin{bmatrix} \mathbf{x}(k+1) \\ v(k+1) \end{bmatrix} = \begin{bmatrix} \mathbf{A}_d - \mathbf{b}_d \mathbf{k}^T & \mathbf{b}_d k_i \\ -\mathbf{c}^T T & 1 \end{bmatrix} \begin{bmatrix} \mathbf{x}(k) \\ v(k) \end{bmatrix} + \begin{bmatrix} 0 \\ T \end{bmatrix} r(k).$$

The closed-loop characteristic polynomial of the augmented system is given as:

$$\Delta_a(z) = \begin{vmatrix} z\mathbf{I} - \mathbf{A}_d + \mathbf{b}_d^T \mathbf{k}^T & \mathbf{b}_d k_i \\ \mathbf{c}^T T & z \end{vmatrix},$$

where $\mathbf{I}$ is an identity matrix of order $n$.

Next, we may choose a desired characteristic polynomial of $(n+1)$ order, and perform pole placement design for the augmented system. The location of the integrator pole in the complex $z$-plane may be selected by trial and error, keeping in view the desired settling time of the closed-loop system. This is illustrated in the following example.

**Example 8.18:** The DC motor model.

The discrete-time model of a small DC motor is given as ($T = 0.02s$):

$$\begin{bmatrix} i_{k+1} \\ \omega_{k+1} \end{bmatrix} = \begin{bmatrix} 0.134 & -0.038 \\ 0.038 & 0.816 \end{bmatrix} \begin{bmatrix} i_k \\ \omega_k \end{bmatrix} + \begin{bmatrix} 0.863 \\ 0.053 \end{bmatrix} V_k \quad y_k = \begin{bmatrix} 0 & 1 \end{bmatrix} \begin{bmatrix} i_k \\ \omega_k \end{bmatrix}.$$

The control law for the tracking PI controller is given as:

$$u_k = -k_1 i_k - k_2 \omega_k + k_i v_k,$$

where $v_k = v_{k-1} + T(r_k - \omega_k)$ represents the integrator output.

The augmented system model for the pole placement design using integral control is given as:

$$\begin{bmatrix} i_{k+1} \\ \omega_{k+1} \\ v_{k+1} \end{bmatrix} = \begin{bmatrix} 0.134 & -0.038 & 0 \\ 0.038 & 0.816 & 0 \\ 0 & -0.02 & 1 \end{bmatrix} \begin{bmatrix} i_k \\ \omega_k \\ v_{k+1} \end{bmatrix} + \begin{bmatrix} 0.863 \\ 0.053 \\ 0 \end{bmatrix} u_k + \begin{bmatrix} 0 \\ 0 \\ 0.02 \end{bmatrix} r_k.$$

The desired $s$-plane pole locations for the augmented system model are selected as: $s = -50, -50 \pm j50$. The corresponding $z$-plane pole locations ($T = 0.02s$) are obtained as: $z = e^{-1}, e^{-1 \pm j1}$. The resulting controller gains, obtained using the MATLAB "place" command, are given as: $k_1 = 0.43, k_2 = 15.44, k_i = -297.79$. An update rule for the controller is obtained as:

$$v_k = v_{k-1} - 0.02 (r_k - \omega_k)$$
$$u_k = -0.43 i_k - 15.44 \omega_k + 297.8 v_k$$

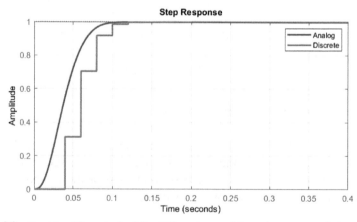

**Figure 8.9**  Tracking PI control of the DC motor model: analog system design; discrete system design.

The step response of the closed-loop system is plotted below (Figure 8.9). The step response of the continuous-time system is plotted alongside. As seen from the figure, the output in both cases attains the reference value of unity in about 0.1 sec.

The MATLAB script for this example is given below:

```
A=[-100 -5;5 -10]; B=[100;0]; C=[0 1]; % define dc motor model
Aa=[A [0;0];-C 0]; Ba=[B;0];        % augmented system model
T=.02;                               % sampling time
[Ad,Bd]=c2d(A,B,T);                  % discretize dc motor model
Ad=[Ad [0;0];-.02*C 1]; Bd=[Bd;0]; % DT augmented sys model
Ka=place(Aa,Ba, -50+[0 50j -50j]); % pole placement for CT system
Kd=place(Ad,Bd, exp(-1+[0 j -j])); % pole placement for DT system
Tss= ss(Aa-Ba*Ka,[0;0;1],[0 1 0],0); % CT closed-loop system
Td=ss(Ad-Bd*Kd,[0;0;T],[0 1 0],0,T); % DT closed-loop system
step(Tss,Td), grid                   % plot step response
legend('Analog','Discrete')          % figure legend
```

## Skill Assessment Questions

Link to the answers:
http://www.riverpublishers.com/book_details.php?book_id=449

1. The linearized model of an inverted pendulum is described as: $\frac{d}{dt}\begin{bmatrix} \theta \\ \omega \end{bmatrix} = \begin{bmatrix} 0 & 1 \\ \frac{g}{l} & 0 \end{bmatrix}\begin{bmatrix} \theta \\ \omega \end{bmatrix} + \begin{bmatrix} 0 \\ 1 \end{bmatrix} u$, where $l = 1$ m, $g = 9.8\frac{m}{s^2}$. The closed-loop system is desired to achieves ITAE specifications for $\omega_n = 5\frac{rad}{s}$.

   (a) Use the Bass-Gura formula to design a state feedback controller for the model.

   (b) Use the Ackermann's formula to design a state feedback controller for the model.

   (c) Use the Sylvester's equation to design a state feedback controller for the model.

   (d) Use the MATLAB "place" command to design a state feedback controller for the model.

2. The model of a small armature-controlled DC motor is given as:
$$\frac{d}{dt}\begin{bmatrix} i_a \\ \omega \end{bmatrix} = \begin{bmatrix} -R/L & -k_b/L \\ k_t/J & -b/J \end{bmatrix}\begin{bmatrix} i_a \\ \omega \end{bmatrix} + \begin{bmatrix} 1/L \\ 0 \end{bmatrix} V_a.$$

   Assume the following parameter values: $R = 1, L = .01, J = .01, b = .1, k_t = k_b = 0.02$.

   (a) Design a state feedback controller for the motor to achieve: $t_s \leq 20$ ms, $\zeta \geq 0.8$.

   (b) Plot the step response of the closed-loop system.

   (c) Design a tracking PI controller to track a constant reference input.

   (d) Plot the step response of the closed-loop system.

3. The simplified model of airplane longitudinal dynamics using angle of attack $(\alpha)$ and pitch rate $(q)$ as state variables is described as: $\frac{d}{dt}\begin{bmatrix} \alpha \\ q \end{bmatrix} = \begin{bmatrix} \frac{Z_\alpha}{V} & 1 \\ M_\alpha & M_q \end{bmatrix}\begin{bmatrix} \alpha \\ q \end{bmatrix} + \begin{bmatrix} \frac{Z_E}{V} \\ M_E \end{bmatrix}\delta_E.$

   Assume the following parameter values: $\frac{Z_\alpha}{V} = -1, \frac{Z_E}{V} = -.1, M_\alpha = -7, M_q = -.5, M_E = -5$.

   (a) Design a full state feedback controller to achieve: $\omega_n \geq 7\frac{rad}{s}$, $\zeta \geq 0.7$.

(b) Plot the step response of the closed-loop system.

(c) Design a tracking PI controller to track the angle of attack reference.

(d) Plot the step response of the closed-loop system.

4. An inverted pendulum over cart model is described using cart position $(y)$ and pendulum angle $(\theta)$ by the following equations: $(M + m)\ddot{y} + ml\ddot{\theta} = f$, $m\ddot{y} + ml\ddot{\theta} - mg\theta = 0$.

Assume the following parameter values: $m = 1$ kg; $l = 1$ m; $M = 10$ kg; $g = 9.8$.

(a) Design a state feedback controller to achieve: $t_s \leq 1s$, $\%OS \leq 2\%$.

(b) Plot the step response for $\theta(t)$ and $y(t)$.

5. The simplified model of a flexible beam is given as: $\dfrac{d}{dt}\begin{bmatrix} x \\ v \end{bmatrix} =$

$\begin{bmatrix} 0 & 1 \\ -\omega_n^2 & -2\zeta\omega_n \end{bmatrix}\begin{bmatrix} x \\ v \end{bmatrix} + \begin{bmatrix} 0 \\ \omega_n \end{bmatrix} u$, where $\omega_n^2 = 10^4$, $\zeta = .005$.

(a) Design a state feedback controller to achieve the following specifications: $t_s \leq 20$ ms, $\zeta \geq 0.8$.

(b) Design a tracking PI controller for the beam to achieve the following specifications: $\omega_n \geq 100\frac{\text{rad}}{\text{s}}$, $\%OS \leq 5\%$, $e(\infty)|_{step} = 0$.

6. Consider the model of human postural dynamics in the sagittal plane, described as an inverted pendulum, given as: $G(s) = \dfrac{k}{s^2-\Omega^2}$, where $k = 0.01$, $\Omega = 3$ rad/ sec.

(a) Obtain a state variable model of postural dynamics.

(b) Design an pole placement controller to achieve: $\omega_n \geq 5\frac{\text{rad}}{\text{s}}$, $\xi \geq 0.85$.

(c) Design a tracking PI controller to track the postural command $\theta_{ref}$.

(d) Plot the step response of the closed-loop system.

7. Consider the model of an inverted pendulum described in Q1 above.

(a) Choose a sampling time $T$ and discretize the state variable model.

(b) Design a pole placement controller for the discrete model.

(c) Compare the step response of analog and discrete systems.

8. Consider the model of a DC motor described in Q2 above.

(a) Choose a sampling time $T$ and discretize the state variable model.

(b) Design a pole placement controller for the discrete model.

(c) Compare the step response of analog and discrete systems.

9. Consider the mode of aircraft longitudinal dynamics described in Q3 above.

   (a) Choose a sampling time $T$ and discretize the state variable model.
   (b) Design a pole placement controller for the discrete model.
   (c) Compare the step response of analog and discrete systems.

10. Consider the model of a flexible beam described in Q5 above.

   (a) Choose a sampling time $T$ and discretize the state variable model.
   (b) Design a pole placement controller for the discrete model.
   (c) Compare the step response of analog and discrete systems.

11. Consider the model of human postural dynamics described in Q6 above.

   (a) Choose a sampling time $T$ and discretize the state variable model.
   (b) Design a pole placement controller for the discrete model.
   (c) Compare the step response of analog and discrete systems.

12. Consider the model of inverted pendulum over cart described in Q4 above.

   (a) Choose a sampling time $T$ and discretize the state variable model.
   (b) Design a pole placement controller for the discrete model.
   (c) Compare the step response of analog and discrete systems.

# 9

# Frequency Response Design
## of Compensators

## Learning Objectives

1. Characterize the frequency response of transfer function models.
2. State the performance criteria for controller design in the frequency domain.
3. Design a variety of compensators for dynamic system models using frequency response methods.
4. Characterize the closed-loop frequency response of the compensated system.

In this final chapter, we discuss the frequency response methods for designing compensators for feedback control systems. Frequency response methods predate the root locus method and the time-domain (state variable) design methods. These methods base the controller design on the frequency response of the loop transfer function, which maybe empirically obtained.

The frequency response of a system can be graphed in multiple ways. The two most common representations are: (1) the Bode plot and (2) the Nyquist or Polar plot. The Bode magnitude and phase plots are traditionally used to design controllers; the Nyquist plot is used to determine the stability of the closed-loop system. The closed-loop frequency response may be visualized on the Nichol's chart.

The frequency response design seeks to impart a certain degree of relative stability to the control loop, as measured by gain and phase margins determined on the frequency response plots. The closed-loop stability is alternatively ascertained via the Nyquist criterion that relates to the encirclements of the $-1 + j0$ point on the Nyquist plot by the frequency response curve.

Control systems design via the frequency response method requires knowledge of the frequency response of the loop transfer function $KGH(j\omega)$. Strictly speaking, the knowledge of the plant transfer function $G(s)$ is not needed. In fact, in the absence of a mathematical model, the plant transfer function can be identified from empirical measurements of the frequency response $G(j\omega)$.

The peak gain in the closed-loop frequency response is a measure of the relative stability; the higher the peak, the lower the relative stability. The frequency at which the resonant peak occurs, or the system bandwidth if the closed-loop frequency response does not exhibits a peak, is a measure of speed of response in the time-domain.

In the following, we first review the plotting of frequency response and the associated performance metrics. Later, we will discuss the frequency response modification by adding phase-lead, phase-lag, lead–lag, PD, PI, and PID compensators to the control loop.

## 9.1 Frequency Response Representation

We consider standard feedback control system configuration (Figure 5.1), where the feedback loop includes a plant $G(s)$, a sensor $H(s)$, and a controller $K(s)$. Unless stated otherwise, a unity-gain feedback ($H(s) = 1$) is assumed.

The frequency response of the loop transfer function, $KGH(s)$, is given as: $KGH(j\omega) = KGH(s)|_{s=j\omega}$.

We note that for a particular value of $\omega$, the $KGH(j\omega)$ is a complex number, which may be described in terms of its magnitude and phase as $KGH(j\omega) = |KGH(j\omega)|e^{j\phi(\omega)}$.

As $\omega$ varies from 0 to $\infty$, $KGH(j\omega)$ can be plotted in the complex plane (the polar plot). Alternatively, both magnitude and phase can be plotted as functions of $\omega$ (the Bode magnitude and phase plots). These two methods are described below.

### 9.1.1 The Bode Plot

For computing and graphing the frequency response, it is convenient to express the loop transfer function as:

$$KGH(s) = \frac{K \prod_{i=1}^{m} \left(1 + \frac{s}{z_i}\right)}{s^{n_0} \prod_{i=1}^{n_1} \left(1 + \frac{s}{p_i}\right) \prod_{i=1}^{n_2} \left(1 + 2\zeta_i \frac{s}{\omega_{n,i}} + \frac{s^2}{\omega_{n,i}^2}\right)}.$$

In the above, $m$ is the number of finite zeros, which are assumed to be real-valued, $n_0$ is the number of poles at the origin, $n_1$ is the number of real poles, $n_2$ is the number of complex pole pairs, and $K$ represents the dc gain of the loop transfer function. The associated frequency response function is given as:

$$KGH(j\omega) = \frac{K \prod_{i=1}^{m} \left(1 + j\frac{\omega}{z_i}\right)}{(j\omega)^{n_0} \prod_{i=1}^{n_1} \left(1 + j\frac{\omega}{p_i}\right) \prod_{i=1}^{n_2} \left(1 - \frac{\omega^2}{\omega_{n,i}^2} + j2\zeta_i\frac{\omega}{\omega_{n,i}}\right)}.$$

It is customary to plot the Bode magnitude plot on log scale as $|G(j\omega)|_{\text{dB}} = 20 \log_{10} |G(j\omega)|$. The magnitude of the frequency response function is represented in dB as:

$$|KGH(j\omega)|_{\text{dB}} = 20 \log K + \sum_{i=1}^{m} 20 \log \left|1 + \frac{j\omega}{z_i}\right| - (20n_0) \log \omega$$

$$- \sum_{i=1}^{n_1} 20 \log \left|1 + \frac{j\omega}{p_i}\right|$$

$$- \sum_{i=1}^{n_2} 20 \log \left|1 - \frac{\omega^2}{\omega_{n,i}^2} + j2\zeta_i\frac{\omega}{\omega_{n,i}}\right|.$$

The slope of the Bode magnitude plot for large $\omega$ is: $-20(n - m)$ dB/decade, where $n - m = n_0 + n_1 + 2n_2 - m$ represents the pole excess of the loop transfer function.

The phase angle $\phi(\omega)$ is plotted in degrees, which is computed as:

$$\phi(\omega) = \sum_{i=1}^{m} \angle \left(1 + \frac{j\omega}{z_i}\right) - n_0(90°) - \sum_{i=1}^{n_1} \angle \left(1 + \frac{j\omega}{p_i}\right)$$

$$- \sum_{i=1}^{n_1} \angle \left(1 - \frac{\omega^2}{\omega_{n,i}^2} + j2\zeta_i\frac{\omega}{\omega_{n,i}}\right).$$

The phase angle for large $\omega$ is given as: $\phi(\omega) = -90°(n - m)$.

The Bode plot of a higher order system is a composition of Bode plots of first and second order factors representing the real and complex poles and zeros. The composite plot can be obtained by superposition.

**Obtaining Bode Plot in MATLAB.** In the MATLAB Control Systems Toolbox, the Bode plot is obtained using the "bode" command, invoked after defining the transfer function as follows:

```
G=tf(num,den);              % define transfer function
bode(G),grid                % obtain Bode plot
```

### 9.1.2 The Nyquist Plot

For a particular value of $\omega$, the frequency response function, $KGH(j\omega)$, represents a complex number with magnitude and phase, where the latter restricted to $[0, 2\pi]$. The locus of all such points, as $\omega$ varies from $0 \rightarrow \infty$, constitutes the polar plot of $KGH(j\omega)$.

The shape of the polar plot depends on the number and locations of the poles and zeros of $KGH(j\omega)$. It can be visualized by computing its magnitude and phase at selected frequencies (typically low and high frequencies).

For example, if $KGH(s)$ has no poles at the origin, then at low frequencies, $KGH(j0) \cong K\angle 0°$, and at high frequencies, $|KGH(j\infty)| \rightarrow 0$, $\angle KGH(j\infty) = -90°(n - m)$, where $n - m$ represents the pole excess of $KGH(s)$. The resulting polar plot starts from the value of $K$ on the real axis and terminates at the origin approaching it from an angle $-90°(n - m)$.

The Nyquist plot of $KGH(s)$ includes the polar plot of $KGH(j\omega)$ as $\omega$ varies from $-\infty$ to $\infty$; the polar plot of $KGH(j\omega)$ for the negative frequencies is a reflection about the real-axis of the positive frequency plot. Hence, Nyquist plot is a closed curve.

The celebrated Nyquist stability criterion states that the number of unstable closed-loop poles of $\Delta(s) = 1 + KGH(s)$ equals the number of unstable open-loop poles of $KGH(s)$ plus the number of CW encirclements of the $-1 + j0$ point in the complex plane by the Nyquist plot of $KGH(s)$.

**Obtaining Nyquist Plot in MATLAB.** In the MATLAB Control Systems Toolbox the Nyquist plot is obtained by using the "nyquist" command, invoked after defining the transfer function as follows:

```
G=tf(num,den);              % define transfer function
nyquist(G),grid             % obtain Nyquist plot
```

The following examples illustrate the shapes of the Bode and Nyquist plots for the first-, second-, and third-order transfer functions.

**Example 9.1:** A first-order transfer function.

Let $G(s) = \frac{1}{s+1}$; then, $KG(j\omega) = \frac{K}{1+j\omega} = \frac{K}{\sqrt{1+\omega^2}} \angle -\tan^{-1}\omega$

Low frequency: $G(j0) = K \angle 0°$
High frequency: $G(j\infty) = 0 \angle -90°$

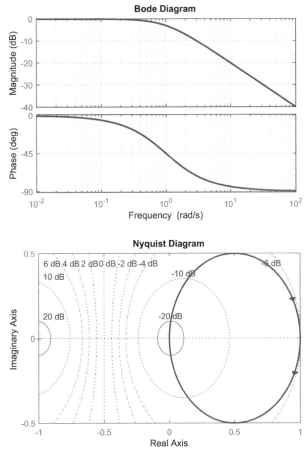

**Figure 9.1** Bode and Nyquist plots for $G(s) = \frac{1}{(s+1)}$.

The Bode magnitude plot has a value of 0 dB (for $K = 1$) and a slope of zero at low frequencies. The slope changes to $-20$ dB/decade at high frequencies and is represented by a straight line sloping downwards on the Bode magnitude plot (Figure 9.1). These two asymptotes meet at the corner frequency ($\omega_0 = 1$). The phase plot shows a variation from $0°$ to $-90°$ with a phase of $-45°$ at the corner frequency.

The polar plot describes a semi-circle in the fourth quadrant (Figure 9.1). The Nyquist plot is a closed circle that includes the reflection of the polar plot for negative $\omega$. As the Nyquist plot stays away from the $-1 + j0$ point for positive $K$, the closed-loop system is stable for all $K > 0$.

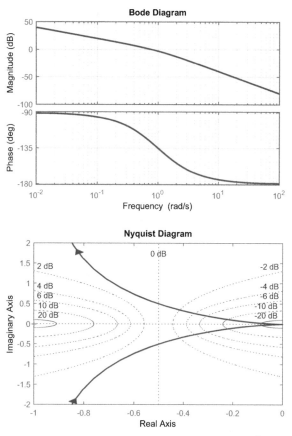

**Figure 9.2**   Bode and Nyquist plots for $G(s) = \frac{1}{s(s+1)}$.

**Example 9.2:** A second-order transfer function with a pole at the origin.

Let $G(s) = \frac{1}{s(s+1)}$; then, $KG(j\omega) = \frac{K}{j\omega(1+j\omega)} = \frac{K}{\omega}\frac{1}{\sqrt{1+\omega^2}}\angle -90° - \tan^{-1}\omega$

Low frequency: $G(j0) = \infty \angle -90°$
High frequency: $G(j\infty) = 0 \angle -180°$

The Bode magnitude plot has an initial slope of $-20$ dB/decade due to the pole at the origin. The slope changes to $-40$ dB/decade at the corner frequency. The phase plot shows a variation from $-90°$ to $-180°$ with a phase of $-135°$ at the corner frequency (Figure 9.2).

The polar plot describes a curve that begins along negative $j\omega$-axis and approaches the origin along negative real axis in the third quadrant

(Figure 9.2). The Nyquist plot includes the reflection of the polar plot for negative $\omega$. As the Nyquist plot stays away from the $-1+j0$ point for positive $K$, the closed-loop system is stable for all $K > 0$.

**Example 9.3:** A third-order transfer function.

Let

$$G(s) = \frac{2}{s(s+1)(s+2)};$$

then,

$$KG(j\omega) = \frac{2K}{j\omega(1+j\omega)(2+j\omega)}$$

$$= \frac{K}{\omega}\frac{1}{\sqrt{1+\omega^2}}\frac{2}{\sqrt{2+\omega^2}}\angle -90° - \tan^{-1}\omega - \tan^{-1}2\omega.$$

Low frequency: $G(j0) = \infty \angle -90°$
High frequency: $G(j\infty) = 0 \angle -270°$

The Bode magnitude plot has an initial slope of $-20$ dB/decade that changes to $-40$ dB/decade and then to $-60$ dB/decade at the two corner frequencies. The phase plot shows a variation from $-90°$ to $-270°$ (Figure 9.3).

The polar plot is a curve that begins along the negative $j\omega$-axis, crosses the negative real-axis, and approaches the origin along the positive $j\omega$-axis (Figure 9.3). The Nyquist plot includes the reflection of the polar plot for negative $\omega$.

For $K = 1$, the Nyquist plot intersects the $-$ve real-axis at $\omega = \sqrt{2}$. Further, $G(j\sqrt{2}) = -0.34$. As $K$ increases, the Nyquist plot will expand and will encircle the $-1 + j0$ point for $K > 3$; hence, the closed-loop system is stable for $K < 3$.

The MATLAB script for the above examples is given below:

```
% first-order system
G=tf(1,[1 1]);                  % define transfer function
figure(1),bode(G),grid          % obtain Bode plot
figure(2),nyquist(G),grid       % obtain Nyquist plot
% second-order system (pole at the origin)
G=tf(1,[1 1 0]);                % define transfer function
figure(1),bode(G),grid          % obtain Bode plot
figure(2),nyquist(G),grid       % obtain Nyquist plot
% third-order system (pole at the origin)
G=tf(1,[1 3 2 0]);              % define transfer function
figure(1),bode(G),grid          % obtain Bode plot
figure(2),nyquist(G),grid       % obtain Nyquist plot
```

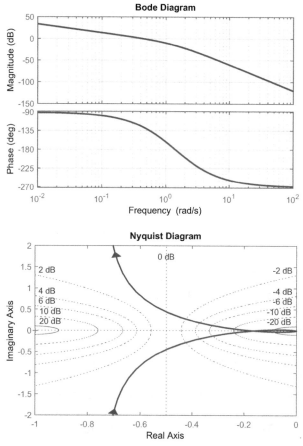

**Figure 9.3**   Bode and Nyquist plots for $G(s) = \frac{1}{s(s+1)(s+2)}$.

## 9.2 Measures of Performance

The measures of performance used with frequency response design methods include gain and phase margins, system error constants, sensitivity and complementary sensitivity functions, etc. These are described next.

### 9.2.1 Relative Stability

In the frequency response design method, the relative stability of the feedback loop is described in terms of the gain and phase margins (Section 5.1.3). We assume that the loop transfer function $KGH(j\omega)$ is stable and minimum-phase, that is, its poles and zeros are located in the left half-plane (LHP) including the $j\omega$-axis. Then, the gain and phase margins are defined as follows.

**Gain Margin.** The maximum amount of loop gain that can be added to the feedback loop (by the controller) without compromising stability.

**Phase Margin.** The maximum amount of phase that can be added to the feedback loop (by the controller) without compromising stability.

The gain margin (GM) on the Bode magnitude plot is indicated as: $\text{GM} = -|G(j\omega_{pc})|_{\text{dB}}$, where $\omega_{pc}$ is the phase crossover frequency, that is, the frequency at which $\phi(\omega_{pc}) = -180°$.

The phase margin (PM) on the Bode phase plot is indicated as: $\text{PM} = 180° + \phi(\omega_{gc})$, where $\omega_{gc}$ is the gain crossover frequency, that is, the frequency at which the Bode magnitude plot crosses the 0 dB line, or attains $|G(j\omega_{gc})| = 1$.

Alternatively, the gain and phase margins can be obtained from the polar plot, which enters the unit circle at the gain crossover frequency $\omega_{gc}$ and crosses the negative real-axis at the phase crossover frequency $\omega_{pc}$.

In particular, assuming that the polar plot crosses the negative real-axis at $g \angle -180°$, the gain margin is given as: $\text{GM} = g^{-1}$.

Similarly, assuming that the polar plot enter the unit circle at $1 \angle -\phi$, then the phase margin is given as: $\text{PM} = 180° - \phi$.

For low-order stable minimum-phase loop transfer functions with number of poles ($n \leq 2$), the polar plot does not cross the negative real-axis; hence, the $GM = \infty$.

A plant transfer function with pole excess ($n - m \geq 3$) has a finite gain margin $0 < GM < \infty$. In such cases, if the polar plot crosses the negative real-axis to the left of $-1$ point, then $GM < 0$ and the closed-loop system is unstable. For $n - m = 2$, the Nyquist plot may cross the negative real-axis, indicating a finite gain margin.

Further, the gain margin equals the maximum controller gain on the root locus plot before the closed-loop roots cross the $j\omega$-axis to migrate to the right half-plane (RHP). For $n \leq 2$, the root locus plot is restricted to the LHP; whereas, for $n - m \geq 3$, it crosses the stability boundary into the RHP for large $K$ whose value can be determined from the magnitude condition.

**Return Difference.** In the case of stable minimum-phase loop transfer functions, the closed-loop stability depends on the polar plot keeping a finite distance from the $-1+j0$ point. The minimum distance of the polar plot from the $-1 + j0$ point, given by the minimum value of $\max_\omega |1 + KGH(j\omega)|$, is called the minimum return difference. The minimum return difference is a measure of the robustness of the control loop to parameter variations in the model.

**Relative Stability Determination in MATLAB.** The "margin" command in the MATLAB Control Systems Toolbox is used to obtain the GM and PM as well as the gain and phase crossover frequencies on the Bode plot. The command is invoked after defining the loop transfer function, as follows:

```
G=tf(num,den);          % define transfer function
margin(G),grid          % obtain stability margins
```

**Example 9.4:** A second-order system model.

Let $G(s) = \frac{1}{s(s+1)}$; then, from the Bode plot, GM $= \infty$ and PM $= 51.8°$ (Figure 9.4(a)).

**Example 9.5:** A third-order system model.

Let $G(s) = \frac{2}{s(s+1)(s+2)}$; then, from the Bode plot we have, GM $=$ 15.6 dB and PM $= 53.4°$ (Figure 9.4(b)).

### 9.2.2 Phase Margin and the Transient Response

The phase margin of a minimum-phase transfer function is related to the damping ratio of the dominant poles of the closed-loop system transfer function. In order to characterize this relationship, we may consider a prototype second-order system with loop transfer function given as:

$$KGH(j\omega) = \frac{\omega_n^2}{j\omega\,(j\omega + 2\zeta\omega_n)}$$

Then, for $|KGH(j\omega_{gc})| = 1$, we have $\omega_n^2 = \omega_{gc}\sqrt{\omega_{gc}^2 + (2\zeta\omega_n)^2}$, which reveals: $\omega_{gc} = k\omega_n$, where $k = \sqrt{\sqrt{4\zeta^4 + 1} - 2\zeta^2}$. The resulting phase margin is computed as: $\phi_m = \tan^{-1}\left(\frac{2\zeta}{k}\right)$.

In particular, in the range of $0.5 < \zeta < 0.9$, we have, $k \cong \frac{3.34-2\zeta}{3}$; hence, $\phi_m \cong \tan^{-1}\left(\frac{6\zeta}{3.34-2\zeta}\right)$.

The phase margin may be approximated as: $\phi_m = 100\zeta$, for $\zeta \le 0.6$. Furthermore, in order to have a $\zeta = 0.7$, we may design the system for a $PM \cong 66°$.

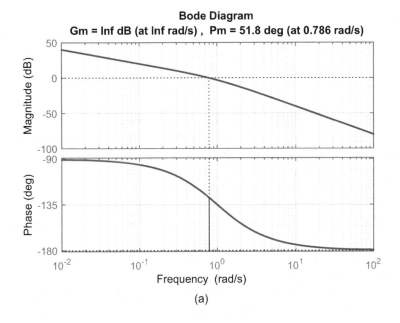

(a)

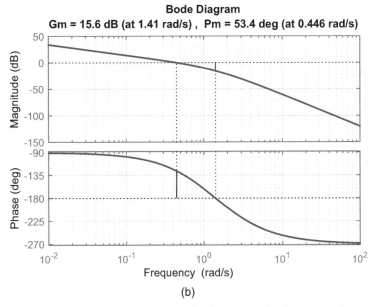

(b)

**Figure 9.4** The relative stability from the Bode plot: a second-order transfer function (a); a third-order transfer function (b).

**The Transient Response.** The phase margin is also related to the settling time of the closed-loop system response. In particular, for a second order transfer function, the settling time $t_s$ and the phase margin $\phi_m$ are related as:

$$\tan \phi_m = \frac{2\zeta\omega_n}{\omega_{gc}} = \frac{8}{t_s\omega_{gc}}$$

**Cloud-Loop Bandwidth.** The transient response is related to the closed-loop bandwidth $\omega_B$ of the system. Assuming that the closed-loop frequency response, $T(j\omega)$, has a unity gain at low frequencies, that is, $T(j0) = 0$ dB, $\omega_B$ is defined at the lowest frequency where $|T(j\omega_B)| = -3$ dB.

The bandwidth obeys the following bounds: $\omega_n \le \omega_B \le 2\omega_n$. For systems with low damping, $\omega_B \cong \omega_n$.

The bandwidth is related to the rise time $t_r$ of the closed-loop system step response as: $\omega_B t_r \approx 1$ or $t_r \cong 1/\omega_B$.

In particular, in the case of second-order systems ($0.6 < \zeta < 0.9$), the rise time can be approximated as: $t_r \cong \frac{3\zeta}{\omega_n}$.

The Bode's gain-phase relation (see book references) defines a unique relation between the Bode magnitude and phase plots of a minimum-phase loop transfer function. In particular, to obtain a good PM, control system designers aim for a moderate slope of $-1$ at the gain crossover frequency, $\omega_{gc}$.

The following example illustrates the relationship between the relative stability and phase margin.

**Example 9.6:** DC motor model.

We consider the model of a small DC motor, given as: $G(s) = \frac{500}{s^2+110s+1025}$. Let a PI compensator for the model be defined as: $K(s) = \frac{K(s+10)}{s}$.

We may select $K = 10$, which results in the closed-loop roots located at: $s = -50 \pm j50.4$. The Bode plot of the loop gain for the compensated system (Figure 9.5(a)) displays a phase margin of $\phi_m = 65.8°$, which corresponds to a damping ratio of $\zeta = 0.7$.

The step response of the compensated system (Figure 9.5(b)) displays a rise time of $t_r = 0.028s$ and a settling time of $t_s = 0.077s$. The predicted value of the settling time is: $t_s = \frac{8}{\omega_{gc}\tan\phi_m} = 0.078s$.

The MATLAB script for this example is given below:

```
G=tf(500,[1 110 1025]);      % define TF model
Kpi=tf([1 10],[1 0]);        % define PI controller
margin(10*Kpi*G)             % Bode plot with stability margins
T=feedback(10*Kpi*G,1)       % closed-loop system
step(T), grid                % step response of compensated sys
```

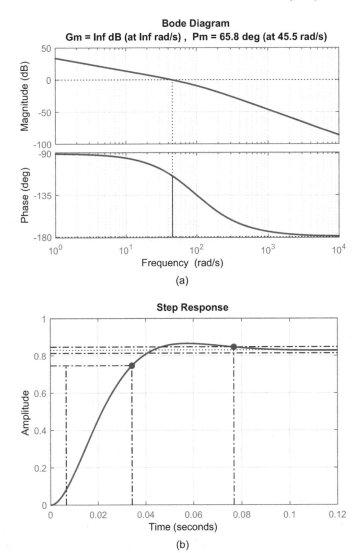

**Figure 9.5** The DC motor model with PI controller: Bode plot of the loop transfer function (a); step response of the compensated system (b).

### 9.2.3 Error Constants and System Type

The system error constants (Section 5.3) are used to compute the steady-state tracking error to a step or a ramp input. The position and velocity error constants are defined as:

$$K_p = \lim_{s \to 0} KG(s)$$
$$K_v = \lim_{s \to 0} sKG(s).$$

These constants can be inferred from the Bode plot at follows:

The position error constant is given by the low frequency asymptote on the Bode magnitude plot, that is, $K_p = \lim_{s \to 0} |KG(j\omega)|$.

The velocity error constant is given as the slope of the low frequency asymptote on the Bode magnitude plot, that is, $\lim_{s \to 0} |KG(j\omega)| = \frac{K_v}{\omega}$.

The system type is defined by the slope of the Bode magnitude plot in the low frequency region. Thus, a slope of 0 dB implies a type 0 system, whereas a slope of $-1$, that is, $-20$ dB/decade implies a type 1 system, etc. Further, a type 1 system indicates the presence of an integrator in the feedback look, which implies a zero steady-state error to a step reference input.

### 9.2.4 System Sensitivity

The sensitivity of the closed-loop transfer function, $T(j\omega)$, to changes in the plant transfer function, $G(j\omega)$, is a function of frequency, $\omega$, and is given as:

$$S_G^T(j\omega) = \frac{\partial T/T}{\partial G/G} = \frac{1}{1 + KGH(j\omega)}$$

Further, sensitivity and the complementary sensitivity, the latter represented by the closed-loop transfer function, are fundamentally constrained as:

$$S(j\omega) + T(j\omega) = 1.$$

In order for the sensitivity to be small over a frequency band, the loop gain $KGH(j\omega)$ must be large. In general, increasing the loop gain decreases stability margins, that is, a trade-off between the sensitivity and relative stability would need to be found.

Accordingly, the control system designers generally aim for high loop gain at low frequencies to obtain $T(j\omega) \cong 1$. The loop gain is reduced at high frequencies to raise the stability margins.

Loopshaping is a graphical technique that aims to impart a desired loop shape to the plot of $KGH(j\omega)$ by adding compensator poles and zeros to the feedback loop (see book references).

## 9.3 Frequency Response Design

The Bode plot serves as a main design tool for the frequency response design method in control systems. The performance indices used to evaluate the frequency response design are the GM and the PM of the loop transfer function $KGH(j\omega)$.

To ensure stability of the closed-loop system, both the GM and the PM must be positive. Additionally, the PM should be adequate to ensure good relative stability and acceptable transient response. The GM should be adequate to achieve good robustness and low sensitivity to parameter variations.

In the following, we assume that the plant and the compensator transfer functions are given by $G(s)$ and $K(s)$, respectively. The characteristic equation of the closed-loop system is given as:

$$\Delta(s) = 1 + KGH(s) = 0.$$

We assume that the plant transfer function has pole excess $(n - m \geq 3)$ so that $0 < \text{GM} < \infty$. Further, a unity-gain feedback $(H(s) = 1)$ is assumed.

In the frequency response design method, the choice of compensators includes the gain compensator, phase-lag and phase-lead compensators, and PD, PI, and PID compensators. These are described next.

### 9.3.1 Gain Compensation

The gain compensation involves choosing a static compensator: $K(s) = K$, which raises the loop gain by $K$, that is, the Bode magnitude plot is shifted up by $20 \log_{10} K$. The resulting change in $\omega_{gc}$ affects the gain crossover frequency and the PM.

In particular, let $\omega'_{gc}$ denote the new gain crossover frequency; then, the new phase margin is given as: $\text{PM} = 180° - \phi(\omega'_{gc})$.

The phase margin is reduced for $(K > 1)$, and increased for $(K < 1)$. Further, the system bandwidth increases for $(K > 1)$, which improves the transient response by reducing the settling time. This is illustrated in the following example.

**Example 9.7:** Let $G(s) = \frac{2}{s(s+1)(s+2)}$; the available gain and phase margins, found from the MATLAB 'margin' command, are given as: GM $= 3$ (9.54 dB), $\omega_{pc} = 1.41\frac{\text{rad}}{\text{s}}$; PM $= 32.6°, \omega_{gc} = 0.749\frac{\text{rad}}{\text{s}}$.

We aim to increase the phase margin to $50°$ using gain compensation. From the Bode plot, we find that: $\phi(j0.49) = -130°$ and $|G(j0.49)| = 5$ dB.

Thus, the required gain $K$ to achieve a $PM = 50°$ is given as: $-5$ dB or $K = 0.56$.

We may note that the loop gain was reduced in order to increase the PM. The design is verified by plotting the frequency response for the loop transfer function: $KG(s) = \frac{1.12}{s(s+1)(s+2)}$, which shows $PM = 50°$ at the new $\omega_{gc} = 0.49 \frac{rad}{s}$ (Figure 9.6).

The MATLAB commands for the gain compensator example are given as:

```
G=tf(2,[1 3 2 0]);        % define transfer function
margin(G),grid            % obtain stability margins
margin(0.56*G),grid       % increased stability margins
```

### 9.3.2 Phase-Lag Compensation

In the frequency response design method, the phase-lag compensator serves dual purpose: it improves the phase margin (a measure of transient response), as well as improves the DC gain (a measure of steady-state response). The phase-lag compensator is described by the transfer function:

$$K(s) = \frac{K(1 + s/\omega_z)}{1 + s/\omega_p}, \quad K(j\omega) = \frac{K(1 + j\omega/\omega_z)}{(1 + j\omega/\omega_p)}, \quad \omega_z > \omega_p.$$

Assuming $K = 1$, the phase-lag compensator response at low and high frequencies is given as:

Low frequency: $K(j0) = 0$ dB $\angle 0°$
High frequency ($\omega \to \infty$): $K(j\omega) = -20 \log_{10}(\frac{\omega_z}{\omega_p})\angle 0°$.

Since the high-frequency gain of the phase-lag compensator is less than unity, the addition of the compensator to the feedback loop lowers the gain crossover frequency and increases the phase margin.

Alternatively, let $K = \omega_z/\omega_p$; then, the compensator gain at low and high frequencies is given as:

Low frequency: $G(j0) = 20 \log_{10} K \angle 0°$
High frequency ($\omega \to \infty$): $G(j\omega) = 0$ dB$\angle 0°$

Since the DC gain of the compensator is greater than unity, addition of the compensator to the feedback loop will boost the relevant error constant by a factor of $K = \omega_z/\omega_p$.

Due to the destabilizing effect of the phase-lag, the compensator pole and zero locations for steady-state error improvement are selected in accordance with: $\omega_p < \omega_z < 0.1\omega_{gc}$, where $\omega_{gc}$ is the gain crossover frequency of the loop transfer function.

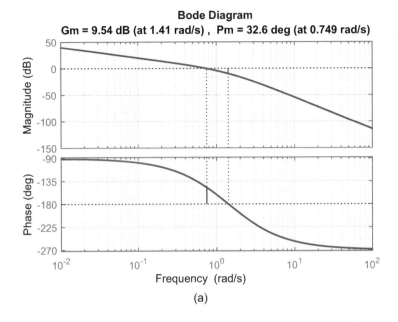

(a)

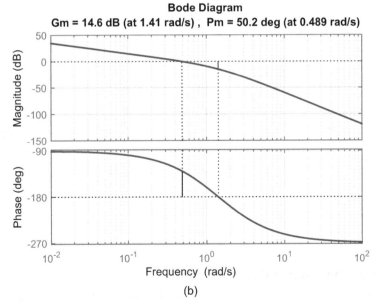

(b)

**Figure 9.6** Gain compensation for the desired phase margin improvement: uncompensated system (a); compensated system (b).

Assuming that the compensated system is desired to have a phase margin of $\phi_m$, the phase-lag compensator design follows these steps:

1. Adjust the DC gain $K$ for the loop transfer function $KGH(s)$ to satisfy the low-frequency gain requirements.
2. On the Bode plot of $KGH(j\omega)$, select the frequency $\omega_1$ so that $\angle KGH(j\omega_1) = -180° + \phi_m + 5°$; then, $\omega_1$ serves as the new gain crossover frequency. The $5°$ safety margin serves to compensate for the estimated: $\angle K(j\omega_1) \cong -5°$.
3. Select pole and zero frequencies as:

$$\omega_z = 0.1\omega_1, \quad \omega_p = \frac{\omega_z}{|KGH(j\omega_1)|}.$$

4. Draw the Bode plot for the compensated system and verify that the design requirements are met.

The phase-lag compensator design is illustrated using the following example.

**Example 9.8:** Let $G(s) = \frac{2}{s(s+1)(s+2)}$; PM $= 32.6°$. The design specifications are: PM $= 50°$, and $e_{ss}|_{ramp} < 0.1$, that is, $K_v > 10$. Then, the phase-lag compensator design steps are:

1. Choose $KG(s) = \frac{22}{s(s+1)(s+2)}$ for $K_v = 11$ to meet the $e_{ss}$ requirement. Draw the Bode plot for $KG(j\omega)$.
2. From the Bode plot, choose $\omega_1 = 0.4$ rad/s, to obtain $\angle KG(j\omega_1) = -123°$; then, $|KG(j\omega_1)| = 24.9$ (28 dB)
3. Choose $\omega_z = 0.04$, $\omega_p = 0.0016$; then, $K(s) = \frac{11(1+s/0.04)}{1+s/0.0016}$
4. The Bode plot of $KG(j\omega)$ has $\omega_{gc} = 0.4$ rad/s and a phase margin of PM $= 51°$ (Figure 9.7).

The MATLAB commands for the phase-lag design example are given as:

```
G=tf(2,[1 3 2 0]);              % define transfer function
bode(11*G),grid,                % draw Bode plot
K= tf([1/.04 1],[1/.0016 1])    % define compensator
margin(11*G*K),grid, hold       % obtain stability margins
```

### 9.3.3 Phase-Lead Compensation

In the frequency response design method, the phase-lead compensator serves to improve the closed-loop bandwidth, leading to transient response improvements. The phase-lead compensator is described by the transfer function:

$$K(s) = \frac{K(1+s/\omega_z)}{1+s/\omega_p}, \quad K(j\omega) = \frac{K(1+j\omega/\omega_z)}{(1+j\omega/\omega_p)}, \quad \omega_z < \omega_p$$

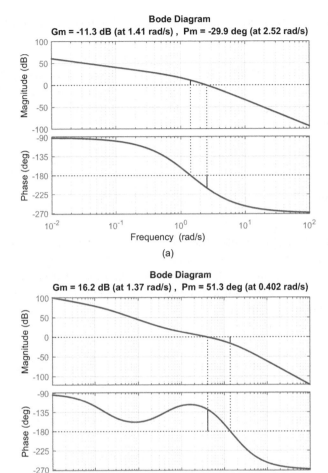

**Figure 9.7** Phase-lag compensation: magnitude compensated plot (a); phase-lag compensated plot (b).

Assuming $K = 1$, the compensator response at low and high frequencies is given as:

Low frequency: $K(j0) = 0 \text{ dB} \angle 0°$

High frequency ($\omega \to \infty$): $K(j\omega) = 20 \log_{10} \left( \frac{\omega_p}{\omega_z} \right) \angle 0°$.

Since the high frequency gain of the compensator is greater than one, the addition of the compensator to the feedback loop increases the gain crossover frequency hence, increasing the bandwidth. The phase-lead compensator thus acts as a high pass filter.

The compensator contributes maximum phase lead, $\theta_m$, at a frequency $\omega_m = \sqrt{\omega_z \omega_p}$. Hence, for maximum effectiveness, the pole and zero of the phase-lead compensator should be placed in the vicinity of the gain crossover frequency $\omega_{gc}$. The compensator transfer function at $\omega_m$ is given as:

$$G_c(\omega_m) = \sqrt{\frac{\omega_p}{\omega_z}} \angle \theta_m, \quad \tan \theta_m = \frac{1}{2}\left(\sqrt{\frac{\omega_p}{\omega_z}} + \sqrt{\frac{\omega_z}{\omega_p}}\right).$$

Practically, the maximum achievable value of $\theta_m$ is about $70°$ that corresponds to $\sqrt{\frac{\omega_p}{\omega_z}} \cong 6$.

The phase-lead design starts by selecting a desired gain crossover frequency $\omega_{gc}$ to meet the desired bandwidth and/or the settling-time requirement. Let the desired settling-time be $t_s$, then the desired gain crossover is, $\omega_{gc} = \frac{8}{t_s \tan \phi_m}$. In addition the phase contribution from the compensator should be positive, that is:

$$\theta = -180° + \phi_m - \angle GH(j\omega_{gc}) > 0°.$$

The loop gain at the desired gain crossover frequency is constrained as: $KGH(j\omega_{gc}) = 1\angle\theta$, which is expanded to obtain:

$$\left(1 + j\frac{\omega_{gc}}{\omega_z}\right)|GH(j\omega_{gc})| = \left(1 + j\frac{\omega_{gc}}{\omega_p}\right)(\cos\theta + j\,\sin\theta).$$

When separated into real and imaginary parts, the above equation can be solved for $\omega_z$ and $\omega_p$, which gives:

$$\omega_p = \frac{\sin\theta}{\cos\theta - |GH|}\omega_{gc}, \quad \omega_z = \frac{|GH|\sin\theta}{1 - |GH|\cos\theta}\omega_{gc}.$$

The above equations are used to compute the pole and zero locations of the phase-lead compensator.

**Simplified Phase-Lead Design.** A simplified method for the phase-lead design proceeds as follows: let $\alpha = \frac{\omega_p}{\omega_z}$; and define $\sin\theta_m = \frac{\alpha-1}{\alpha+1}$, so that $\alpha = \frac{1+\sin\theta_m}{1-\sin\theta_m}$. Since $\omega_m = \sqrt{\omega_z \omega_p}$, we choose the following locations for compensator pole and zero:

$$\omega_z = \frac{\omega_m}{\sqrt{\alpha}}; \quad \omega_p = \omega_m\sqrt{\alpha}.$$

Further, the gain of the compensator at crossover is $K(\omega_{gc}) = \frac{K}{\sqrt{\alpha}}$. Hence, in order to ensure $|KGH(\omega_{gc})| = 1$, we choose the compensator gain as: $K = \frac{\sqrt{\alpha}}{|GH(\omega_{gc})|}$.

The phase-lead design (using both regular and simplified methods) involves the following steps:

1. Choose a desired crossover frequency $\omega_{gc}$ as the greater of $\frac{8}{t_s \tan \phi_m}$ and $\omega_1$ where $\angle KGH(j\omega_1) = -180° + \phi_m + 5°$.
2. Compute compensator phase $\theta = \phi_m - \angle G(j\omega_{gc}) - 180°$; compute $|GH(j\omega_{gc})|$.
3. Solve for $\omega_z$ and $\omega_p$ using the regular or the simplified method.
4. Inspect the Bode plot of the compensated system and verify that the design requirements have been met.

The phase-lead design is illustrated using the following example.

**Example 9.9:** Let $G(s) = \frac{2}{s(s+1)(s+2)}$; PM $= 32.6°$. The design specifications are: $\phi_m = 50°$, $t_s = 4$ s. Then, the phase-lead compensator design steps are:

1. From the settling time requirement, $\omega_{gc} = \frac{2}{\tan \phi_m} = 1.68 \ \frac{\text{rad}}{\text{s}}$; from the Bode plot, $G(j1.68) = 0.23 \angle -189°$.
2. Compute $\theta = 50° + 9° = 59° \cong 60°$.
3. The solution to the regular phase-lead compensator design equations gives: $\omega_z = 0.38 \ \frac{\text{rad}}{\text{s}}, \omega_p = 5.43 \ \frac{\text{rad}}{\text{s}}, K(s) = \frac{14.29(s+0.38)}{s+5.43}$. The Bode plot of the compensated system has $\phi_m = 50.4°$ at $\omega_{gc} = 1.69 \ \frac{\text{rad}}{\text{s}}$ (Figure 9.8).
4. The simplified method with $(\theta_m = 60°, \alpha \cong 13.92)$ returns the compensator: $K(s) = \frac{16(s+0.45)}{s+6.26}$. The Bode plot of the compensated system has $\phi_m = 50.8°$ at $\omega_{gc} = 1.68 \ \frac{\text{rad}}{\text{s}}$.

The MATLAB commands for the phase-lead design example are:

```
G=tf(2,[1 3 2 0]);          % define transfer function
bode(G),grid, hold          % draw Bode plot
K= tf([1/.38 1],[1/5.43 1]) % define compensator
margin(G*K),grid, hold      % obtain stability margins
```

### 9.3.4 Lead-Lag Compensation

A lead-lag compensator combines the phase-lead and phase-lag sections; the phase-lead section improves the bandwidth and the phase margin, and the

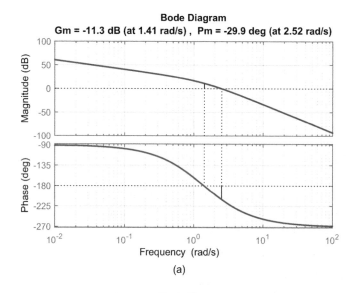

(a)

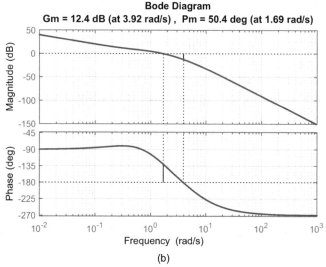

(b)

**Figure 9.8**    Phase-lead compensation: uncompensated plot (a); phase-lead compensated (b).

phase-lag section improves the phase margin and the DC gain. Since both lead and lag sections can contribute to the phase margin improvement, the desired PM improvement can be distributed among the two sections. The lead-lag compensator transfer function is given as: $K(s) = K_{lead}(s)K_{lag}(s)$.

The steps to design a lead-lag compensator are as follows:

1. Choose a static gain $K$ to meet the steady-state error requirement.
2. Design the phase-lag section to meet part of the phase margin requirement.
3. Design phase-lead section to meet the bandwidth/settling time requirement.

The lead-lag compensator design is illustrated using the following example.

**Example 9.10:** Let $G(s) = \frac{2}{s(s+1)(s+2)}$; PM $= 32.6°$. The design specifications are: $\phi_m = 50°$, $t_s = 4s$, and $e_{ss}|_{ramp} < 0.1$. The design steps for the lag-lead design are as follows:

1. Choose $K = 11$ to meet the $e_{ss}$ requirement. Draw the Bode plot for $KG(s)$. The plot has $\omega_{gc} = 2.5$ rad/s and PM $= -30°$.
2. **Phase-Lag Design**. In the phase-lag design, we aim to raise the phase margin to PM $\approx 40°$. From the plot, let $\omega_1 = 0.5$ rad/s, such that KG $(j0.5) = 18.6$ (25.8 dB)$\angle -130°$.
3. Complete the phase-lag design with: $\omega_z = 0.05$, $\omega_p = 0.003$.
4. Draw the Bode plot for $K_{lag}G(j\omega)$. The plot shows a PM of $40°$.
5. **Phase-Lead Design.** Choose $\omega_{gc} = 1.7$ rad/s to meet the $t_s$ requirement. Then, from the Bode magnitude plot, $K_{lag}G(j1.7) = 0.15$ ($-16.5$ dB)$\angle -191°$.
6. Compute the required compensator phase angle $\theta = 62°$, and solve the phase-lead compensator equations to get: $\omega_z = 0.24$, $\omega_p = 4.7$.
7. The lead-lag compensator is described as: $K(s) = \frac{11(s+0.05)(s+0.24)}{(s+0.003)(s+4.7)}$. The Bode plot for the compensated system has $PM = 50.2°$ at $\omega = 1.71$ rad/s (Figure 9.9).

The MATLAB commands for the lead-lag design example are:

```
G=tf(2,[1 3 2 0]);              % define transfer function
bode(11*G),grid, hold           % draw Bode plot
K1= tf([1/.05 1],[1/.003 1])    % define lag compensator
margin(11*G*K1),grid,           % obtain stability margins
K2= tf([1/.24 1],[1/4.71 1])    % define lead compensator
margin(11*G*K1*K2),grid, hold   % obtain stability margins
```

### 9.3.5 PI Compensator

The PI compensator adds an integrator to the feedback loop, and is used for steady-state error improvement. The PI compensator transfer function is

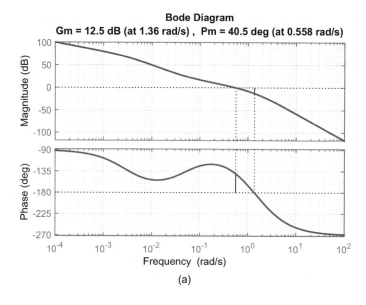

(a)

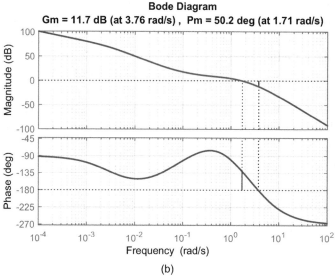

(b)

**Figure 9.9** Lag-lead compensation: phase lag compensated system (a); lag-lead compensated system (b).

defined as:

$$K(s) = k_p + \frac{k_i}{s}, \ K(j\omega) = \frac{k_i(1 + j\omega/\omega_z)}{j\omega}, \quad \text{where } \omega_z = \frac{k_i}{k_p}.$$

The PI compensator is normally designed for a high frequency gain of one, i.e., $k_p = 1$. Additional gain compensation, as in the case of phase-lag design, can be used to increase the phase margin to some desired value. The zero location for the PI compensator is arbitrarily selected close to the origin as shown in the following example.

**Example 9.11:** Let $G(s) = \frac{2}{s(s+1)(s+2)}$; PM $= 32.6°$. The design specifications are: PM $= 50°$, and $e_{ss}|_{ramp} < 0.1$, i.e., $K_v > 10$. The PI compensator design steps are as follows:

1. For PM $= 50°$, we have $-180° + \phi_m + 5° = -125°$. The Bode plot for $G(j\omega)$ has, for $\omega = 0.42\frac{rad}{s}$, $G(j0.42) = 2.11 \ (6.6 \text{ dB})\angle -125°$. Thus, a gain compensation of $-6.6$ dB is needed.
2. For the PI design, let $\omega_z = 0.04$, $K(s) = \frac{s+.04}{s}$.
3. The Bode plot of $0.465 \ KG(j\omega)$ has $\omega_{gc} = 0.42$ rad/s and a phase margin of PM $= 49.8°$ (Figure 9.10).

The MATLAB commands for the PI design example are:

```
G=tf(2,[1 3 2 0]);            % define transfer function
bode(G),grid, hold            % draw Bode plot
Kpi= tf([1 .04],[1 0])        % define PI compensator
margin(0.465*G*Kpi),grid,     % obtain stability margins
```

### 9.3.6 PD Compensator

The PD compensator adds a first-order zero to the numerator of the loop transfer function, which increases the bandwidth hence improves the transient response. The PD compensator is described as:

$$K(s) = k_d s + k_p, \ K(j\omega) = k_p(1 + j\omega/\omega_z)$$

The desired gain crossover $\omega_{gc}$ is selected as the greater of $\frac{8}{t_s \tan \phi_m}$ and $\omega_1$, where $\angle GH(j\omega_1) = -180° + \phi_m - 90° + 5°$; the $90°$ term in this expression refers to the phase added by the PD compensator, and $5°$ is safety margin.

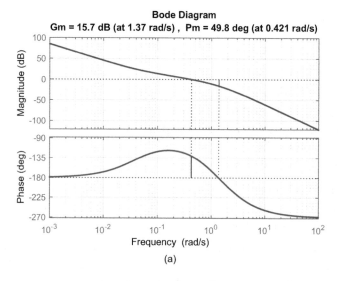

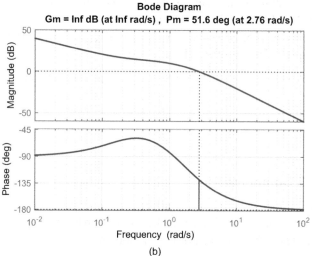

**Figure 9.10**    PI compensation (a); PD compensation (b).

Let $\omega_{gc}$ denote the desired gain crossover frequency; then, from the requirement: $|KGH(j\omega_{gc})| = 1$, the zero location is selected as:

$$\omega_z = \frac{|GH|}{\sqrt{1 - |GH|^2}}\omega_{gc}.$$

The PD compensator design is illustrated using the following example.

**Example 9.12:** Let $G(s) = \frac{2}{s(s+1)(s+2)}$; PM $= 32.6°$. The design specifications are: $\phi_m = 50°$, $t_s = 4s$.

1. From the settling time requirement, $\omega'_{gc} = \frac{2}{\tan\phi_m} \cong 1.7\,\frac{\text{rad}}{\text{s}}$; $G(j1.7) = 0.23\,\angle -190°$. From the angle requirement, $\omega_1 = 2.8\,\frac{\text{rad}}{\text{s}}$; $G(j2.8) = 0.07\,\angle -215°$.
2. From the design equation, using $\omega_1$, the compensator zero location is given as: $\omega_z = 0.2\,\frac{\text{rad}}{\text{s}}$, $K(s) = (s+0.2)$.
3. The Bode plot of the compensated system has $\phi_m = 51.6°$ at $\omega_{gc} = 2.76\,\frac{\text{rad}}{\text{s}}$ (Figure 9.10).

The MATLAB commands for the phase-lead design example are:

```
G=tf(2,[1 3 2 0]);        % define transfer function
bode(G),grid, hold        % draw Bode plot
Kpd= tf([1 .2],1)         % define PD compensator
margin(5*G*Kpd),grid,     % obtain stability margins
```

### 9.3.7 PID Compensator

The PID compensator represents a combination of the PD and PI designs. As a rule, PI section is designed first, and can be used to realize a part of or the entire phase margin improvement desired. The bode plot of the PI compensated system is plotted, and followed by the PD design, which imparts the desired transient response improvement. The PID compensator design is illustrated using the following example.

**Example 9.13:** Let $G(s) = \frac{2}{s(s+1)(s+2)}$; PM $= 32.6°$. The design specifications are: $\phi_m = 50°$, $t_s = 4$ s.

1. The Bode plot for $G(j\omega)$ has $G(j0.8) = 0.9\,\angle -150°$.
2. **PI design:** We aim for $\phi_m = 30°$ in the PI section. Accordingly, let $\omega_z = 0.05$, $K_{PI}(s) = 0.9(s+.05)/s$. The Bode plot of $K_{PI}G(j\omega)$ has $\omega_{gc} = 0.7$ rad/s and $G(j0.7) = 1\,\angle -148°$.
   From the settling time requirement, $\omega_{gc} = \frac{2}{\tan\phi_m} \cong 1.7\,\frac{\text{rad}}{\text{s}}$; $G(j1.7) = 0.23\,\angle -190°$. From the phase angle requirement, $\omega_1 = 2.5\,\frac{\text{rad}}{\text{s}}$; and $G(j2.5) = 0.1\,\angle -210°$.
3. **PD design:** let the desired $\omega_{gc} = 2.5\,\frac{\text{rad}}{\text{s}}$; from the design formula, the compensator is zero selected as: $\omega_z = 0.2\,\frac{\text{rad}}{\text{s}}$, $K_{PD}(s) = 5(s+0.2)$.

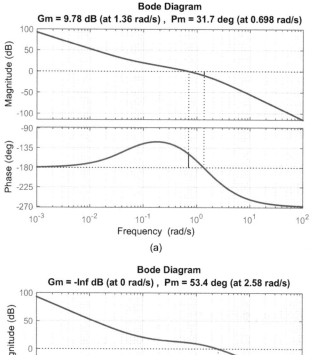

(a)

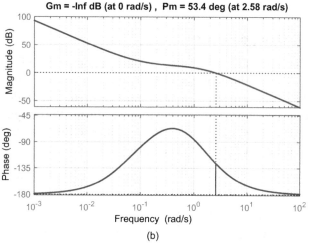

(b)

**Figure 9.11** PID compensator design: PI compensation (a), followed by PD compensation (b).

4. The PID compensator is given as: $K(s) = 4.5(s+0.05)(s+0.2)/s$. The Bode plot of the compensated system has $\phi_m = 53.4°$ at $\omega_{gc} = 2.58\ \frac{\text{rad}}{\text{s}}$ (Figure 9.11).

The MATLAB code for the PID design is given below:

```
G=tf(2,[1 3 2 0]);              % define transfer function
bode(G),grid, hold              % draw Bode plot
Kpi=.9*tf([1 .05],[1 0])        % define PI compensator
margin(G*Kpi),grid,             % obtain stability margins
Kpd=5*tf([1 .2],1)              % define PD compensator
margin(G*Kpi*Kpd),grid,         % obtain stability margins
```

## 9.3.8 Compensator Designs Compared

In this section, we aim to compare the frequency response compensator designs presented in Examples 9.7–9.13 based on the quality of the unit-step response of the closed-loop system. The compensator comparison in shown in Figure 9.12. The salient features of different compensators are captured in Table 9.1. The phase-lead design includes both the regular (phase-lead 1) and simplified (phase-lead 2) designs.

We make the following observations based on Figure 9.12 and Table 9.1.

1. Almost all compensators were able to meet the phase margin requirement; however, the settling time requirement specified in case of phase-lead, lead-lag, PD and PID compensators was not met. It is so because a second-order model was assumed in deriving the settling time formula, whereas the plant transfer function is third-order.

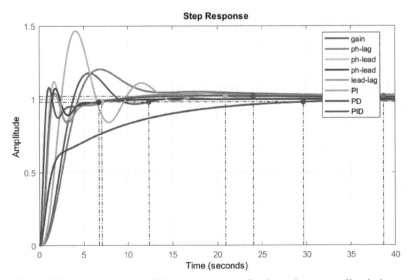

**Figure 9.12** A comparison of the step responses for the various controller designs.

**Table 9.1**   A comparison of the unit step response of the compensated system for six different controller designs for $G(s) = \frac{2}{s(s+1)(s+2)}$

| Compensator Type | Compensator, $K(s)$ | Gain Crossover ($\omega_{gc}$, rad/s) | Phase Margin (PM) | Damping Ratio ($\zeta$) | Percentage Over-shoot | Settling Time |
|---|---|---|---|---|---|---|
| Gain | 0.56 | 0.49 | 50.2° | 0.47 | 18% | 12.4s |
| Phase-lag | $\frac{11(1+s/0.04)}{1+s/0.0016}$ | 0.4 | 50.3° | 0.56 | 21% | 38.7s |
| Phase-lead1 | $\frac{14.29(s+0.38)}{s+5.43}$ | 1.69 | 50.4° | 0.35 | 12% | 7.1s |
| Phase-lead2 | $\frac{13.92(s+0.45)}{s+6.26}$ | 1.5 | 56.6° | 0.41 | 7% | 6.7s |
| Lead-lag | $\frac{11(s+0.05)(s+0.24)}{(s+0.003)(s+4.7)}$ | 1.71 | 50.2° | 0.38 | 3% | 7.2s |
| PI | $\frac{s+0.04}{s}$ | 0.42 | 49.8° | 0.25 | 46% | 20.9s |
| PD | $s+0.2$ | 2.76 | 51.6° | 0.75 | 0% | 29.6s |
| PID | $\frac{4.5(s+0.05)(s+0.2)}{s}$ | 2.58 | 53.4° | 0.43 | 8% | 28s/6.3s |

2. From Figure 9.12, the settling times for phase-lead and lead-lag compensators are around 7s. Though the MATLAB computed settling time for the PID compensator (based on $\pm2\%$ limits) is 28s, the settling time with $\pm2.1\%$ limits is 6.3s.

3. Though 50° degree phase margin is achieved by almost all compensators, the effective closed-loop damping is lower than $\zeta = 0.5$ as seen from the overshoots in the step response. This is again due to the fact that estimates are based on a second-order system model.

## 9.4 Closed-Loop Frequency Response

The closed-loop frequency response, that is, the frequency response of the closed-loop system reveals important information about the relative stability of the feedback loop and the speed of response in the time-domain.

The closed-loop frequency response of a plant described by, $G(j\omega)$, with a static gain controller, $K$, connected in a unity gain feedback configuration ($H(s) = 1$) is computed as:

$$T(j\omega) = \frac{KG(j\omega)}{1 + KG(j\omega)}$$

To obtain a better understanding of the closed-loop frequency response, let $KG(j\omega) = X(j\omega) + jY(j\omega)$; then,

$$T(j\omega) = \frac{X + jY}{(1 + X) + jY} = Me^{j\phi} \quad \text{and}$$

$$|T(j\omega)|^2 = \frac{X^2 + Y^2}{(1 + X)^2 + Y^2} = M^2,$$

The latter equation is rearranged to obtain:

$$\left(X + \frac{M^2}{M^2 - 1}\right)^2 + Y^2 = \frac{M^2}{(M^2 - 1)^2}.$$

The above relation represents the equation of a circle with center at $X = \frac{M^2}{1 - M^2}, Y = 0$, and radius $r = \frac{M}{1 - M^2}$. Hence, the constant magnitude loci for the closed-loop system constitute a series of constant $M$ circles on the frequency response plot of the loop transfer function.

Further, as $M \to 1$, $X = -\frac{1}{2}$, which is a vertical line that separates the circles with $M < 1$ from those with $M > 1$.

The phase relationship similarly reveals an equation for constant phase circles, described as:

$$\left(X + \frac{1}{2}\right)^2 + \left(Y - \frac{1}{2N}\right)^2 = \frac{1}{4} + \left(\frac{1}{2N}\right)^2$$

where $N = \tan\phi$. The constant phase circles are centered at: $X = -\frac{1}{2}$, $Y = \frac{1}{2N}$, with radius: $r = \sqrt{\frac{1}{4} + \left(\frac{1}{2N}\right)^2}$.

**Bandwidth and Resonance Peak.** The closed-loop frequency response typically displays low-pass characteristics. Its two important parameters are the bandwidth and the resonant peak.

The closed-loop bandwidth is defined by: $T(j\omega_b) = \frac{1}{\sqrt{2}}$, or $|T(j\omega_b)|^2 = \frac{1}{2}$.

For a prototype second-order transfer function:

$$T(j\omega) = \left.\frac{\omega_n^2}{s^2 + 2\zeta\omega_n s + \omega_n^2}\right|_{s=j\omega},$$

the bandwidth is given as: $\omega_b = \omega_n\sqrt{1 - 2\zeta^2 + \sqrt{2 - 4\zeta^2 + 4\zeta^4}}$. In particular, $\omega_b = \omega_n$ for $\zeta = 0.7$.

The resonance peak in the frequency response occurs at $\omega = \omega_r$ for $\zeta < 0.7$ and can be obtained by setting the derivative of $|T(j\omega)|$ to zero. The results are:

$$\text{Resonant frequency: } \omega_r = \omega_n\sqrt{1 - 2\zeta^2}$$

$$\text{Resonant peak: } M_r = |T(j\omega_r)| = \frac{1}{2\zeta\sqrt{1-\zeta^2}}$$

When the polar plot of the loop gain, $KGH(j\omega)$, is superimposed onto the constant $M$ and constant $N$ contours, it reveals the magnitude peak $M_r$ in $T(j\omega)$.

The magnitude peak represents a measure of relative stability, while the resonant frequency provides a measure of speed of response in the time-domain. A value of $M_r = 1.3$ (or 2.5 dB) is considered a good compromise between speed and stability.

**The Nyquist Plot.** The MATLAB Control System Toolbox plots the constant $M, N$ contours on the Nyquist plot by turning on the "grid" following the plotting of the frequency response:

```
G=tf(num,den);          % define the transfer function
nyquist(G); grid        % nyquist plot with const M circles
```

**The Nichol's Chart.** An alternate way to visualize the frequency response is to plot it on the Nichol's chart, where the magnitude in dB is plotted along the vertical axis, and the phase in degrees is plotted along the horizontal axis. The "grid" command is similarly used with the Nichol's chart to plot the constant $M, N$ contours to aid in the visualization of the closed-loop frequency response. This is done by invoking the following MATLAB commands:

```
G=tf(num,den);          % define the transfer function
nichols(G); grid        % nichol chart with const M circles
```

The resonance peak in the closed-loop frequency response is illustrated using the following example.

**Example 9.14:** Let $G(s) = \frac{10}{s(s+1.86)}$, $G(j\omega) = \frac{10}{-\omega^2+j1.86\omega}$; then, the closed-loop frequency response has a peak $M_r = 5$ dB, observed on both the Nyquist plot and the Nichol's chart (Figure 9.13). The corresponding resonant frequency can be obtained by clicking on the frequency response plot, which gives: $\omega_r = 2.88$ rad/s.

The corresponding MATLAB commands are:

```
G=tf([10],[1 1.86 0]);  % define the transfer function
nyquist(G),grid         % nyquist plot
nichols(G),grid         % nichol's chart
```

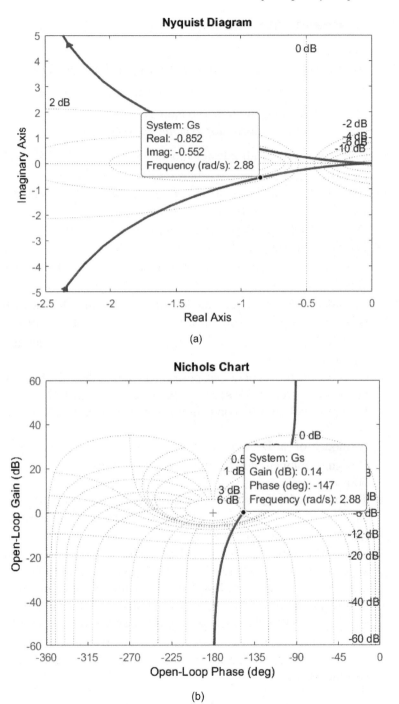

**Figure 9.13**   Closed-loop frequency response on the Nyquist plot (a), and the Nichol's chart (b).

## Skill Assessment Questions

Link to the answers:
http://www.riverpublishers.com/book_details.php?book_id=449

1. Consider a plant with a transfer function: $G(s) = \frac{10}{s^2+2s+10}$.
   (a) Use gain compensation to obtain a $\phi_m \geq 65°$.
   (b) Modify the compensator to achieve: $KGH(0) \geq 5$.

2. Consider a plant with transfer function $G(s) = \frac{5(3s+1)}{s(s^2+2s+10)}$. Design a phase-lead compensator to achieve: $t_s \leq 1$ s, $\phi_m \geq 50°$.

3. Consider the model of a DC motor with the following parameter values: $R = 1\ \Omega$, $L = 10$ mH, $J = 0.01$ kgm$^2$, $b = 0.1\ \frac{\text{Ns}}{\text{rad}}$, $k_t = k_b = 0.02$. Design a lead-lag compensator to achieve: $\omega_B \geq 100\ \frac{\text{rad}}{\text{s}}$, $\phi_m \geq 45°$.

4. Consider the simplified model of a flexible beam, given as: $G(s) = \frac{10^4}{s^2+10s+10^4}$. Design a lead-lag compensator to achieve: $\omega_B \geq 100\ \frac{\text{rad}}{\text{s}}$, $\phi_m \geq 45°$.

5. Consider the simplified model of an automobile, described as: $G(s) = \frac{28s+120}{s^2+7s+14}$. Design a lead-lag compensator to achieve: $\omega_B \geq 10\ \frac{\text{rad}}{\text{s}}$, $K_p = \infty$, $K_v = 20$, $M_p \leq 1$ dB. State the gain and phase margins achieved.

6. Consider the model of human postural dynamics described as an inverted pendulum, given as: $G(s) = \frac{k}{s^2-\Omega^2}$, where $k = 0.01, \Omega = 3$ rad/sec. Design a lead-lag/PID compensator for the model to achieve: $t_s \leq 2$ s, $\phi_m \geq 50°$.

7. A model of a DC motor is given as: $G(s) = \frac{1}{(0.1s+1)(0.02s+1)}$. Design a compensator for the model to achieve: $PM \geq 60°, \omega_{gc} \geq 50$ rad/sec, $e_{ss}(\text{step}) \leq 10\%$.

8. The model of the pitch axis of a quad-copter with actuator is described as: $G(s) = \frac{1}{s^2(s+10)}$. Design a Phase-lead controller to achieve a settling time of 4 sec with a phase margin of $\phi_m \geq 60°$.

9. Consider the input-output model of magnetic levitation system, given as: $G(s) = \frac{5K}{s^2+20}$, where $K = 100$. Design a phase-lead controller to achieve a settling time of less than 1 sec with 60° phase margin.

10. A servomechanism is described by the plant transfer function: $G(s) = \frac{10}{s(s+1)(s+10)}$. The closed-loop system is desired to have a settling time less than 0.5 sec, a phase margin of at least 50° and less than 5% error to a ramp input. Design a lead-lag controller to achieve the response specifications.

# Appendix

## The Laplace Transform

The Laplace transform of a causal time function $f(t)$ is a complex function $f(s)$, defined as:

$$\mathcal{L}[f(t)] = f(s) = \int_0^\infty f(t)e^{-st}\,dt,$$

where $s = \sigma + j\omega$ is a complex variable. In case of discontinuities, the lower limit is changed to $0^-$.

Laplace transform of simple functions may be computed by hand. For example:

$$\mathcal{L}[u(t)] = \int_0^\infty e^{-st}\,dt = \frac{1}{s}$$

$$\mathcal{L}[e^{-at}u(t)] = \int_0^\infty e^{-(s+a)t}\,dt = \frac{1}{s+a}$$

$$\mathcal{L}[e^{j\omega t}u(t)] = \int_0^\infty e^{j\omega t}e^{-st}\,dt = \frac{1}{s-j\omega} = \frac{s+j\omega}{s^2+\omega^2}.$$

Laplace transform of complex functions is obtained by invoking its properties:

1. Linearity: $\mathcal{L}[a_1 f_1(t) + a_2 f_2(t)] = a_1 f_1(s) + a_2 f_2(t)$
2. Time shift: $\mathcal{L}[f(t-T)] = e^{-sT}f(s)$
3. Scaling: $\mathcal{L}[f(at)] = f(\frac{s}{a})$
4. Differentiation: $\mathcal{L}[\frac{df(t)}{dt}] = sf(s) - f(0)$
5. Integration: $\mathcal{L}[\int_0^t f(\tau)\,d\tau] = \frac{f(s)}{s}$
6. Initial value theorem: $f(0) = \lim_{s\to\infty} sf(s)$
7. Final value theorem: $f(\infty) = \lim_{s\to 0} sf(s)$
8. Convolution: $\mathcal{L}[f(t) \times g(t)] = F(s)G(s)$

A dynamic system model is often obtained in terms of an ODE. A potent advantage of using Laplace transform as an analysis tool is that it transforms an ODE into an algebraic equation that can be solved by algebraic methods. For example,

$$\frac{dy(t)}{dt} + ay(t) = u(t) \overset{\mathcal{L}}{\rightarrow} (s+a)y(s) = u(s) + y(0).$$

The inverse Laplace transform of a complex function may be obtained by using partial fraction expansion (PFE) followed by the use of Laplace transform tables. For example:

$$\mathcal{L}^{-1}\left[\frac{a}{s(s+a)}\right] = \mathcal{L}^{-1}\left[\frac{1}{s} - \frac{1}{s+a}\right] = (1 - e^{-at})u(t).$$

The Laplace transform is also used to obtain the input-output transfer function of a differential equation model. For example, for a first-order ODE model: $\frac{dy(t)}{dt} + ay(t) = u(t)$, the transfer function is found as: $\frac{y(s)}{u(s)} = \frac{1}{s+a}$. Using the transfer function, the step response of the system is computed as: $y(s) = \frac{1}{s+a}\left(\frac{1}{s}\right)$, or $y(t) = \frac{1}{a}(1 - e^{-at})u(t)$.

## The $\mathcal{Z}$-Transform

The $\mathcal{Z}$-transform of a sequence $x(k)$ is defined as:

$$\mathcal{Z}[x(k)] = X(z) = \sum_{k} = -\infty^{\infty} x(k)z^{-k}$$

where $z$ is a complex variable. To obtain a unique $F(z)$ the region of convergence (ROC) for $\mathcal{Z}$-transform must be specified.

We will assume that the sequence $x(k)$ is obtained by sampling a continuous-time causal signal $x(t) \to x(kT)$, so that the lower summation limit can be taken as $k = 0$. Then, the respective $\mathcal{Z}$ and Laplace transform variables are related as: $z = e^{sT}$.

The $\mathcal{Z}$-transforms of sampled values of common signals are given as:

$$\mathcal{Z}[u(kT)] = \frac{1}{1 - z^{-1}}; \quad |z^{-1}| < 1$$

$$\mathcal{Z}[e^{(-akT)}] = \frac{1}{1 - e^{-aT}z^{-1}}; \quad |e^{-aT}z^{-1}| < 1$$

$$\mathcal{Z}[e^{(j\omega kT)}] = \frac{1}{1 - e^{j\omega T}z^{-1}}; \quad |z^{-1}| < 1$$

The $\mathcal{Z}$-transform properties include:

1. Linearity: $\mathcal{Z}[a_1 f_{1(k)} + a_2 f_2(k)] = a_1 F_1(z) + a_2 F_2(z)$
2. Time-shift: $\mathcal{Z}[f(k - n_0)] = z^{-n_0} F(z)$
3. Scaling: $\mathcal{Z}[a^k f(k)] = F(\frac{z}{a})$
4. Time differentiation: $\mathcal{Z}[f(k) - f(k - 1)] = (1 - z^{-1}) F(z)$
5. Time reversal: $\mathcal{Z}[f(-n)] = F(z^{-1})$
6. Final value theorem: $\lim_{k \to \infty} f(k) = \lim_{z \to 1} (z - 1) F(z)$
7. Convolution: $\mathcal{Z}[f_{1(k)} \times f_{2(k)}] = F_{1(z)} F_{2(z)}$

A discrete-time dynamic system is often modeled with difference equations, which can be conveniently solved using $\mathcal{Z}$-transform (see references for details).

A table relating continuous-time signals to their Laplace and $\mathcal{Z}$-transform equivalents can be found at: http://lpsa.swarthmore.edu/LaplaceZTable/LaplaceZFuncTable.html

# Index

# About the Author

**Kamran Iqbal** obtained his B.E. in Aeronautical Engineering from NED University, his M.S. and Ph.D. in Electrical Engineering from the Ohio State University, followed by postdoctoral work at Northwestern University. He has held academic appointments at College or Aeronautical Engineering, GIK Institute of Engineering Science and Technology, University of California, Riverside, University of California, Irvine, California State University at Fullerton, and University of Arkansas at Little Rock where he currently serves as Professor of Systems Engineering. His research interests include neuromechanics and motor control of human movement, computational intelligence, and biomedical engineering and signal processing. He is a senior member of IEEE (USA), a member of IET (UK), ASEE, IASTED, and Sigma Xi, and past president of Sigma Xi Central Arkansas Chapter. He is a regular contributor to the IEEE System, Man, and Cybernetics and/or Engineering in Medicine and Biology conference.